MW00678086

# CAMEOS
# OLD & NEW

# CAMEOS

# OLD & NEW

ANNA M. MILLER

To Kathy
a good friend and
colleague with thanks for
all your support!
Anna M Miller 1991

VNR Van Nostrand Reinhold
New York

Library of Congress Catalog Card Number 90-47269
ISBN 0-442-00278-5



Printed in the United States of America.

Van Nostrand Reinhold
115 Fifth Avenue
New York, New York 10003

Chapman and Hall
2-6 Boundary Row
London, SE1 8HN, England

Thomas Nelson Australia
102 Dodds Street
South Melbourne 3205
Victoria, Australia

Nelson Canada
1120 Birchmont Road
Scarborough, Ontario MIK 5G4, Canada

16 15 14 13 12 11 10 9 8 7 6 5 4 3 2 1

**Library of Congress Cataloging-in-Publication Data**

Miller, Anna M.
Cameos old and new / by Anna M. Miller.
p. cm.
Includes bibliographical references and index.
ISBN 0-442-00278-5
1. Cameos. I. Title.
NK5720.M55 1991
736'.222—dc20

90-47269
CIP

*To Diana Nicholas and*
*Russell Kemp and*
*the Lizzadro Museum*

# Contents

# Preface

In several years of studying the cameo and its development, I have interviewed connoisseurs and collectors throughout the United States, Europe, and the Middle East. I found that the cameo has apparently enjoyed an almost universal appeal since its introduction some twenty-five hundred years ago. The purposes of my research were to find the Hellenistic roots of the cameo; to uncover little-publicized public cameo collections; to gather firsthand information on the new machine-carving techniques; and to investigate cameo design and style from the eighteenth century to the present day. Unlike some glyptics scholars, I take the view that a bas-relief, carved and raised above its background, is appropriately called a cameo whether it is executed in hardstone, shell, or another material.

C.W. King, author of *Antique Gems and Rings,* took exception to the term *cameos* being used for any material except hardstone. "The small heads and busts cut in full relief from gemstones are not cameos in the strictest sense," said King, "but rather portions of statuettes, the remainder of the figure having been originally completed in the precious metals."

This book is based on well-dated and researched ancient and modern writings about cameos produced in all materials, musings of individual collectors, and my own personal views.

*Cameos Old and New* does not exhaust the subject, but it does fill the need for a one-stop reference with current information about this particular art form. Demand for such a text has become acute in recent years, as cameos enjoy a revival as collectibles. And, as more individuals become acquainted with this exquisite, miniature art form, discovering the cameo's historic appeal as well as aesthetic beauty, cameo collecting will grow. There is considerable material in the market, some contemporary, that collectors are overlooking. Much of it is of a quality that will encourage the beginner and sustain the interest of the more advanced collector. This book describes the rewards of cameo collecting and outlines how to identify and separate earlier gems from modern ones, as well as how to develop a feel for quality. Knowing this will help the beginner avoid two of the major pitfalls in cameo collecting: the purchase of mediocrities and the purchase of fakes and forgeries.

The history of the cameo can be of absorbing interest to the jewelry historian and scholar since it is a subject with close ties to fine art, literature, and world history. The motifs, myths, and legends that were the subjects of ancient Greek, Roman, Renaissance, and later works, as well as the styles in which they were portrayed, are not only expressions of a particular culture, but are keys to decoding and dating cameos.

The tools, materials, and carving processes used by craftsmen through the centuries are discussed. This knowledge helps one to distinguish between cameos of various periods, especially in detecting cameos being mass-produced today by the ultrasonic process.

Discovering and learning about cameos can be likened to embarking on an adventure. These miniature works of art can transport you on an imaginary flight to cultures that were in flower long ago. Searching for cameo treasures can begin right in your own family jewel box, because many people have cameos they either bought for themselves or received from a relative or friend. At the very least, close observation of modern or commercial cameos provides an excellent starting point to learning more about the glyptic arts.

And, while cameo connoisseurship and scholarly expertise may be slow (or swift, according to the dictates of the individual), learning about cameos is certain to open new areas of interest, provide amusement, and offer intellectual stimulation.

# Acknowledgments

This book was written with the help of many knowledgeable people: scholars, carvers, collectors, connoisseurs, museum curators, glyptic-arts dealers, and interested friends. It was a great pleasure to research this subject and visit museums, private collections, cameo dealers, and production workshops. I especially wish to thank my colleagues who participated in this project and were enthusiastic enough about the subject to do field research.

In particular I wish to thank Joanna Angel for her special editorial expertise, unflagging support and enthusiasm, and multilingual capabilities upon which I drew time and again. Special thanks also go to Alison Crane, who sought out the rare books on ancient cameos and loaned—without limitation—priceless tomes from her own library. I am grateful, too, to have had the advice and insight of glyptics expert Derek Content. Along with providing gracious access to his three-thousand-volume glyptic-arts library, I benefited from his wise counsel, which greatly influenced this work.

My thanks also go to: Dr. Keith DeVries, Janice Klein, and Colin Vargas, of the University Museum at the University of Pennsylvania, for their courtesy, interest, and photographic access to cameo collections; Diana Lizzadro Nicholas, Russell Kemp, and John Lizzadro of the Lizzadro Museum, Elmhurst, Illinois, for special access, time, and photographic considerations in the middle of a frigid Chicago winter; Gerhard Becker and Andreas Becker, who generously contributed information and assistance in obtaining cameos from the Deutsches Edelsteinmuseum for photographing that considerably strengthens the contents of this book; and Achim Grimm, who supplied significant information.

Special thanks are due to two staff members of the library at the National Museum in Warsaw, Poland, for supplying information concerning cameos and other gem collections of the king Stanislaw August Poniatowski and his nephew Prince Stanislaw Poniatowski.

A separate thank-you goes to my colleague in Egypt, Connie Nelson, whose resourcefulness afforded the opportunity to examine rare and exquisite glyptic collections in Egypt and Turkey.

A special joy in Italy were the people who gave their time and knowledge so generously: Sandro Sebastianelli, Dante Sebastianelli, Nino Tagliamonte, Antonella Borriello, Gennaro Borriello, Basilio Liverino, and others who made research in Torre del Greco such a delight. I especially want to thank the Italian Trade Commission in California for supplying many helpful facts, contacts, and special information on cameos.

An indication of the overall popularity of cameos is the many private collections shown in this book. I am particularly grateful to Ann Spector, for allowing unlimited access to examining, evaluating, and photographing her remarkable cameo collection, and to Judy Spector, for allowing photography of her extraordinary cameos after the Pompeii frescoes.

Inspiration for many of the ideas in this book were provided by discussions with friends and colleagues, and I want to acknowledge: Myra Waller, Waller Antiques, Canada; Richard Drucker; Ken and Elaine Roberts; John Miller; Susan Eisen; Therese Kienstra; Ellen Epstein; Louis Zara; Mary Desmarteau; Joanna Scandiffio, Butterfield & Butterfield; Gloria Lieberman, Skinner Galleries; William Milne, Dunning's Auction Services; Harmer Johnson; Dr. Joseph Sataloff; Tom R. Paradise; Ute Bernhardt; Elise B. Misiorowski, GIA; Dona Dirlam, GIA; Louise T. Hall; Dr. George Sines, University of California; Dr. James Draper, Metropolitan Museum of Art; William Fones; Fred L. Iusi, president of the American Society of Appraisers; Richard Wolf, Sotheby's; Clyde Treadwell, Sonic-Mills; Mrs. Muir Rogers; Dr. A. Bernhard-Walcher, Kunsthistorisches Museum; Dr. Guitty Azarpay, University of California at Berkeley; A. Ruppenthal; W. K. Ngai; Masashi Furuya; Dr. Charles Peavy, ASA; Elizabeth Hutchinson and Van Edwards for art work.

Finally, thanks to all the personal property appraisers of the American Society of Appraisers, especially gems and jewelry discipline, and colleagues from the Gemological Institute of America, who by their calls and interest were constant reminders of the need for information on this subject; and particularly to Judith Baron, who got me involved in this highly engrossing study with her question, "Why don't you write a book about cameos?"

# CAMEOS OLD & NEW

*C h a p t e r*

# 1

# THE HISTORY AND ROMANCE OF THE GLYPTIC ARTS

Cameos—emperors have commissioned them, kings have acquired them, and at least since the fourth century B.C., collectors have cherished them.

Cameos are works of miniature sculptural art. Their original purpose is lost in history, with the original intent of individual pieces known only to the craftsmen who carved them. What is known is that of the countless ancient examples of the sculptor's and carver's craft, none offers such a unique window through which to view the cultural past. Cameos reveal the manners, customs, philosophies, historic events, and social occasions that have marked our past. The ancient cameos, like fine art and sculpture, were intended as statements. Only in the last few hundred years has their significance eroded as cameos began to depict endless profiles—poorly done for the most part—of vapid females.

Today most cameo connoisseurs believe that cameos originated without any practical purpose other than ornamentation, but a Dutch scholar, Zadoks-Josehus Jitta, seeing their deeper meaning has characterized them as "messages in agate." In ancient times the cameo also served as an amulet, a talisman, a storyboard depicting ethics and morals, a tangible affirmation of one's faith, and in some cases a reflection of one's destiny. At one point the wearing of a cameo portrait of the ruling monarch not only showed one's loyalty to the court but also facilitated a quick audience with the ruler; such a cameo could quite conceivably guarantee safety and favor. Given all this, it seems inappropriate to consign the carving of cameos only to the category of personal adornments and amusements plastered on cups, vessels, crowns, and relics, used to fill treasuries of royalty and the church.

The word *cameo* first appeared in the thirteenth century and was used to distinguish a gemstone carved in relief or raised above the background of a stone from a gemstone being engraved as *intaglio,* in which a design has been incised or engraved *into* the stone (fig. 1-1). The source of the word *cameo* is unclear. It may be of Arabic origin, from *khamea,* meaning amulet. Or it may be connected with the Greek words for jewel: *camaheu* or *camaieul.* According to *Webster's Dictionary,* the English word *cameo* is from the Italian *cammeo,* possibly derived from the Middle French *camaieu,* from the Middle Latin *camahutus* or *camaeus.* Some old documents contain the word spelled *camahutum, chamah,* and *camahieu.* It may have come from the Syriac *chemeia,* meaning charm, or from *camaut,* a camel's hump, which was applied metaphorically to anything prominent and may therefore have been used to identify a gem in relief from an intaglio. What is clear is that *cameo* defines the work that has been done, not the material on which it was done.

Preceding cameos by thousands of years were the seals used for personal documents or household goods, the intaglios. Intaglios are stones engraved with pictures, signatures, or both. Intaglios and cameos are both forms of the *glyptic* arts, glyptic being from the Greek word *glyptos,* meaning to carve. In Latin, cases for collections of engraved stones are called *dactyliothecae,* a word that is found in numerous old and traditional texts on gem engraving.

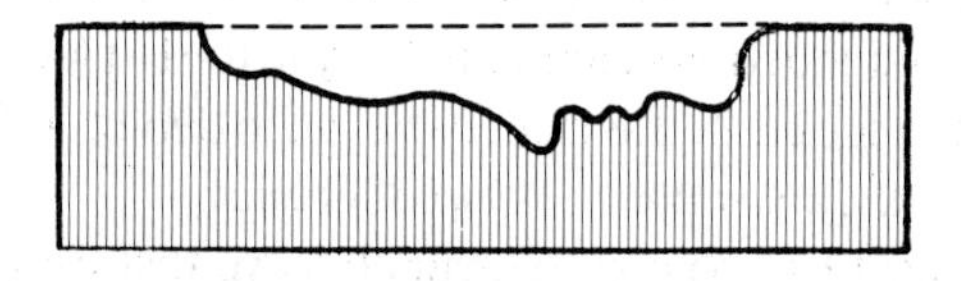

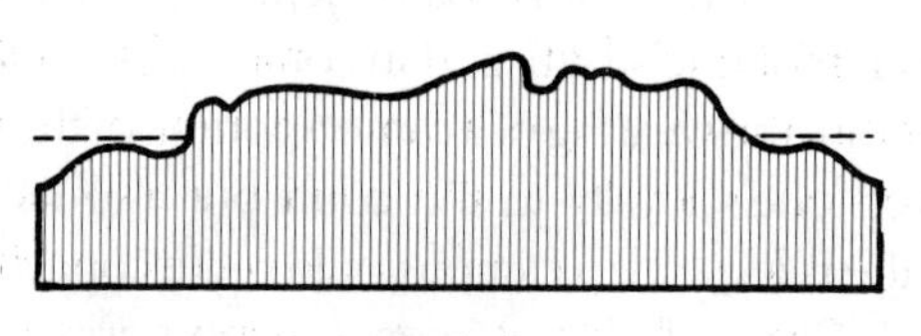

**FIGURE 1-1.**

Top: *In an intaglio the design is engraved or incised into the surface of a stone.* Bottom: *A cameo has the design carved to project above the background of the stone. (Illustration by Elizabeth Hutchinson)*

## THE EARLIEST ENGRAVINGS

The earliest engraved stones—going back at least as far as 15,000 B.C.—were *petroglyphs,* signs or simple figures carved or scratched into rock to record and communicate. The petroglyphs developed into two distinct forms: pictographs, which recorded events in pictures; and ideographs, symbols intended to represent ideas. The early pictographs, or pictorial art, developed into the cuneiform script, the earliest form of writing. (The hieroglyphics of Egypt are pictographs.) All were forerunners of the cylinder seal, predecessor to the stamp seal, scarab, scaraboid, and cameo.

Seals were man's attempt to develop a device that would identify and authorize documents and protect and tax property, as well as serve as an amulet that would guard its wearer according to the strength of his belief. As seals, small engraved stones were used both as insurance against tampering with personal property and as a way to mark property. The seals, made from a variety of materials, including wood, ivory, and stone, had their inscribed design impressed into soft clay or wax that was attached to a letter, cask, jug, jar, or doorway. As long as the seal remained unbroken, the items were perceived to be safely sealed, or locked. Violators of the seals were dealt with harshly, and penalties were swift. When the seal also served as an amulet, it was worn to protect against evil and invite good fortune. Seals have been found in a variety of shapes, including cones and tablet.

In the book *First Impressions,* Dominique Collon wrote that the first important engraved seals were the cylinder seals of Mesopotamia, which can be dated to before 3300 B.C. These seals coincide with the first relief carvings found on the curved surfaces of stone vessels. A great number of these vessels, steles (slabs), and tomb reliefs can be seen today in archaeological museums in Egypt and Turkey. These relief carvings, both high relief and bas-relief, may be closely related to, if not actual predecessors of, the earliest cameos.

The cylinder seal was an elongated hardstone cylinder, usually jasper, serpentine, or limestone, engraved in intaglio and designed for rolling around the openings of jugs or other vessels, imprinting its design into the soft wax or clay (fig. 1-2). A hole was cut through the stone longitudinally, and a string was threaded through, to tie the seal around the owner's wrist or neck. Some of the most ancient seals were produced by the Egyptians who refined the cylinder seal to meet their own needs and finally supplanted it with the scarab seal. The use of the scarab, a dung beetle and Egyptian symbol of eternity, can be traced at least to 3200 B.C.

FIGURE 1-2.

*Cylinder seals typical of the ancient Egyptian period. (From Harold H. Hart, editor,* Jewelry: A Pictorial Archive of Woodcuts and Engravings *[New York, Dover, 1978], p. 14)*

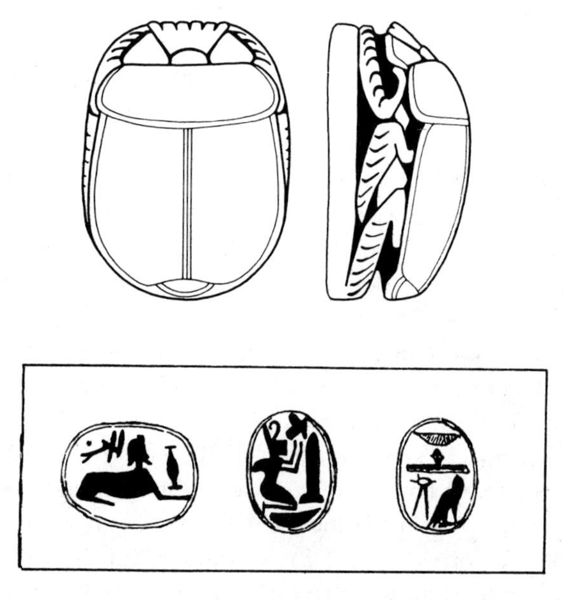

FIGURE 1-3.

*An Egyptian scarab seal with various examples of engraved bases. (From Harold H. Hart, editor,* Jewelry: A Pictorial Archive of Woodcuts and Engravings *[New York, Dover, 1978], p. 15)*

The scarab was half an oval dome, with the engraved flat underside used as a seal (fig. 1-3). The image of the beetle was carved on the domed top of the stone, and the flat base was engraved with symbols of luck and good fortune, prayers for the dead, spells, and names.

The scarab was popular and held in such high regard by its users that hundreds of thousands of them were crafted for the daily use by the populace. Indeed, the scarab was so common that its true meaning was apparently completely misunderstood by the Greek historian Herodotus when he compiled his fifth-century work about the Greco-Persian wars. Herodotus explained the abundance of scarabs by suggesting that they were a type of money, even pronouncing them the "small change" of the Egyptian monetary system.

Ancient Egyptian scarab seals were constructed largely from soft stones such as steatite, a variety of talc often called soapstone; selenite, a translucent colorless form of gypsum with a variety called alabaster; and serpentine, frequently used as a jade substitute. However, the overwhelming majority of Egyptian scarabs were made either from faience, a substance of soapstone and powdered quartz with a fired glaze made from copper compounds, or from an indigenous silica glass. Only when grinding tools found their way into workshops did the craftsmen work the harder materials such as chalcedony, carnelian, and onyx. Egypt was a prolific producer of scarab seals from 3200 to 200 B.C., and introduction of the horizontal shaft on the bow drill, in about 1375 B.C., helped increase and amplify the production. It has been suggested that this improvement over holding the shaft of the bow vertically set the stage for the invention of all lathes.

Scholars believe that the art of engraving was introduced into Assyria from Egypt and later spread to the Mediterranean countries in commerce and trade. Scarabs engraved with both Assyrian and Egyptian symbols have been found in ancient Assyrian digs, so scholars reason that there must have been a very close connection and rich cultural exchange between the two societies. The famed archaeologist Sir Leonard Woolley defined excavation layers in terms of the seals present in his excavations of strata in Ur. And, material sciences professor Dr. George Sines of the University of California states that making seal stones was not a cheap endeavor. According to an 1800 B.C. Assyrian text, "The cost of manufacture was approximately equal to that of an ox and slightly less than that of a slave girl," Sines quotes from the article "Ancient Lapidary," and from personal communications with M. T. Larsen of the University of Copenhagen. While seals had a high ornamental value, their function of ensuring security must have been even more critical to warrant such an expense.

Although the Egyptian craftsmen were productive, there was little poetry in their engraving art. The designs and figures were still, lacking any suggestion of mobility, because craftsmen were compelled by their priests to conform to specific dogma about life and the expression of the models. No craftsman was allowed freedom of expression or license to change the models in any way, the result being that most images were not very lifelike. It was left to the Phoenicians, Greeks, Etruscans, and Romans to expand the art of gem engraving to the point where the seals were considered desirable articles for personal adornment as well as utility.

## Classical Greece: From Scaraboid to Cameo

The Mycenaean civilization in Greece, Crete, and Cyprus (ca. 1600 to 1100 B.C.) produced well-crafted early gold work that included the use of seals in rings. Further, the early cultures learned how to work the harder materials, such as chalcedony and quartz. With the fall of Mycenae in the twelfth century B.C., however, the art of carving gemstones declined and was nearly lost.

The Greeks revived stone engraving sometime between the eleventh and eighth centuries B.C., the Homeric period of Greece, originating a geometrical style of design. The craftsmen, no longer familiar with the wheel and belt drive, were forced to engrave by hand, and the seals and intaglios were crude and simple. The designs, though symbolic, lacked creativity. The Greeks changed the image of the scarab seal to one more representative of their culture with the introduction of the scaraboid. This seal was semi-oval like the scarab but had no beetle image or other ornamentation on the dome. The seal itself was engraved on the flat bottom of the stone.

In the eighth and seventh centuries B.C., gem engraving was affected largely by Eastern art and Oriental culture. Seals of this period were carved chiefly from soft materials such as soapstone and depicted a variety of animal subjects. The Etruscans, who migrated from Asia Minor into what is now central Italy in about 900 B.C., although best known for their gold jewelry work, were also highly skilled in the engraving arts. They were a people of wealth and high technical skill, and they were instructed by the Greeks in gem engraving. Their seal subjects show definite Greek influences, with the popular themes of Heracles, Ajax, Achilles, and a variety of Greek myths predominant. From scenes that adorn the walls of Etruscan tombs, we can see the rich dress of the culture, and their sculpture indicates considerable technical expertise along with talent for realistic portraiture.

The Greeks themselves took much of their mythology and legend from Egypt, Babylonia, and Sumeria. The engravings of the Greek gods and goddesses, along with their personal attributes of power, were carved on stones, which were used not only as amulets to ward off evil and bring good fortune but also were collected as an art form and kind of religious icon. An indication of the genuine antiquity of an engraved stone of this period can be found in the composition: the carving fills the entire field of the stone.

During the sixth-century Archaic period, Greek craftsmen regained the technique of carving hardstones with a drill and cutting wheel. The newly found knowledge resulted in a specific style of cutting, the oval cabochon shape, and so, a flatter style of engraved ringstone.

The engraving began to take on a naturalistic look, while still depicting the legends and myths of the gods and heroes. Engravers made frequent use of the image of Athena, guardian of Athens, the largest city of Hellas, and of the hero Heracles (whom the Romans later called Hercules), a favorite of the Greeks. Also portrayed in stone were an incredible number of centaurs, sileni (woodland deities), maenads (women who participated in Dionysian orgies), and scenes of seduction and rape.

In this period in Greece, seals were widely used in the average household to protect goods and pantries, and the craft of the seal maker was a highly respectable one. It is interesting to note, however, that the engravers were bound by a strict code of law that prohibited them from keeping any records of the designs they made. This was apparently done to discourage the making of reproductions that could be misused by the unethical.

The glyptic arts reached full bloom in the so-called Golden Age of the Greek classical art, from the fifth to the fourth centuries B.C. Engravers and carvers particularly sought the semitranslucent chalcedony, colorful jaspers, reddish-brown carnelians, and both clear and purple quartz for their raw materials. Color and the ability to transmit light were two critical criteria for carving materials: translucency allowed the design of the work to show most effectively, and color produced a more dramatic impact; in addition, colors had their own symbolism, often associated with a particular subject. The multicolored, many-layered sardonyx stone was not well known to the Greeks. Those who had seen it mistook it for an artificial compound of the East. Because of their rich green and blue colors, turquoise and malachite were also desirable materials, and numerous examples of carving are found in these stones.

Before the fifth century B.C., the majority of gem carvers were anonymous, but now artists began to sign their work. A notable craftsman of the time was an Ionian named Dexamenos, a master of classical glyptics. Some of his works have been found in south Russia in the Black Sea region, and a number of his pieces are in the Hermitage collection. Four of the most celebrated glyptics of this artist are: a chalcedony intaglio of a woman seated at her bath facing a mirror that is held in the hand of her handmaiden; a yellow jasper engraved with a crane that is standing on one foot and preening its wing, while under its raised foot a grasshopper sits; an intaglio of a flying crane, often acclaimed as one of the finest engraved works of any artist; and an engraved portrait of a bearded man. All bear the signature of the master Dexamenos.

The Greek carvers depicted a man or animal in a solitary state. And, normally, the Greeks carved their figures nude. They were accomplished at showing tension and straining muscle, with the final effect portraying a man or animal in the prime of life and health.

The use of engraved stones was not restricted in Greece to any particular rank or level of society, as it was in some other cultures. Anyone who wished—and could afford—to was entitled to wear an engraved stone. As the scaraboid had evolved into a ringstone, the finger ring was considered the most convenient vehicle for the signet. Greeks were interested in engraved stones showing a wide variety of subjects, including myths, legends, military victories (more so if the owner-wearer was a warrior), and special events. Of course, the Greeks were highly superstitious and often wore the image of their favorite deity to give them courage and confidence, as well as to affirm their beliefs. The gem carvers and engravers of that time were encouraged in their trade and were rewarded with money and praise.

New gemstone materials in a wide variety of rich colors were introduced into Greece after the eastern campaigns of Alexander the Great. While conquest of territories from the mouth of the Nile to India meant ingress of the Hellenic culture into those far-flung regions, the migration of culture went both ways. The introduction to Greece of new stones from the conquered lands, together with fresh design ideas and accumulations of wealth, set the stage for a new form of carving, the cameo.

## The Cameo Makes Its Appearance

It is widely held by historians that the birthplace of the cameo is Alexandria, the town founded by Alexander the Great at the mouth of the Nile in Egypt in 332 B.C. Cameo expert Gisela M. A. Richter wrote in *Catalogue of*

*Engraved Gems of the Classical Style* that the technical innovation of cameos was clearly derived from the Orient, just one more "instance of the increasing love of luxury which pervaded Greek society in the 3rd and 2nd centuries B.C." It was the introduction of a very special kind of stone, multilayered sardonyx from India and Arabia, that provided the inspiration for the cameo. Instead of the design being cut from, or carved into the stone as it had been for centuries, the cameo was a picture produced in relief, using one or more strata (layers) of the stone in one or more colors. In other words, the cameo produced a picture in the light upper layer of stone, which stood out boldly against the contrasting darker bottom layer. Most cameos only use two layers, but they can have as many as five or more layers of color.

The carvers were able to combine all their skills to arrange their subjects and make use of the multicolored strata of the stones, employing the dramatic zones of color or spotting in those layers to heighten and beautify their pictures. Tools to create the bas-reliefs were simple and already in use: the wheel with belt drive, a set of drills, and abrasives—probably corundum but possibly minute diamond splinters. The potter's wheel, used by the Chinese since the seventh century B.C., had passed into Egypt, Greece, and finally southern Europe. The Etruscans changed the belt drive from a perpendicular wheel to a horizontal spindle. Theodorus of Samos is credited with the invention of the lathe in 600 B.C., and Phoenician engravers were using the drill in 600 B.C. Details of the cameo were executed with iron or bronze gravers.

A new phenomenon in glyptics was the cameo portrait. Alexander commissioned Lysippus to sculpture his portrait, Apelles to paint it, and Pyrgoteles to engrave it in stone. The artists depicted the "Conqueror of the World" using a likeness of Zeus; thus we do not know what Alexander really looked like. Alexander was treated as a god and in fact claimed descent from Hercules and Zeus and kinship with Dionysus, and the early Greek portraits were highly stylized. The features are indistinguishable from those of any god of the same period. During the classical Greek era, only a man of outstanding achievement had a portrait or statue made of himself, and the portrait was set apart from others with a sign or symbol that pointed out the extraordinary exploits of the individual, along with the attributes of the deities. No Greek of the classical period was correctly portrayed on coins; only the heads of deities were depicted.

Alexander's territorial conquests made gold and precious stones available as never before to the average citizen, and many women were able to emulate the goddess of love, bedecking themselves with jewels. For cameo art to survive, however, the society had to be at peace and have time for cultivation of the arts. Moreover, the art of the cameo carver is one that flourishes best in periods of economic prosperity, as Greece enjoyed under Alexander's rule.

---

## Hellenistic Era

The Hellenistic era represented a time in the eastern Mediterranean of an enlarged world for the people of that society, representing a transition from classical Greek to Roman rule.

Variety and diversity are the keynotes of Hellenistic art forms. In gem engraving the subjects are identified to a new degree with nature. Mythological subjects included the old favorites Dionysus and Aphrodite, Eros, Psyche, Artemis, and the Egyptian goddess Isis, and scenes from daily life with idyllic landscapes and pastoral subjects were also depicted. Symbols are also found in the cameos of this period. In general the cameos were formed in oval shapes, and the stones were not particularly thick, they were generally small; the field (or foreground) was almost fully carved, leaving scant contrasting strata of color. Although much cameo work was excellently carved, not all pieces are perfect works of art, and there are numerous examples of inferior workmanship. According to some writers on this subject, a few carvers had only one exclusive talent, usually the ability to execute one particular detail on a subject, such as the drapery on a garment, so that the rest of the cameo remained strictly pedestrian.

During the Hellenistic era, a maiden could do more than pray or use a magic potion to win her love. She could also use her jewelry as a charm; these charms, often cameos, were as powerful in their expression of desire as religious icons. Perhaps Eros dancing on a gem dangling from her bodice was an invitation to a would-be lover. Not surprisingly, jewelry from this era, with both cameos and intaglios, abounds. Many of the cameo pieces were simply miniature sculptures in the round, of very high relief. They were worn in diadems, bracelets, pendants, and as clothing ornaments and were characteristic as people relished luxury and splendor. The bas-reliefs of designs on metal, as well as stones, are more frequent in this period than any earlier Greek era, indicating that artists were interested in three-dimensional representation.

The most remarkable bas-relief work from the Hellenistic period is the world-famous Tazza Farnese, (color plate 1), an 8-inch-wide sardonyx cut in the shape of a shallow dish and believed to have been made in Alexandria. The piece is on view today in the Museo Nazionale in Naples. Produced by an unknown artisan, the vessel is believed to have been made for a second-century Ptolemaic queen. One of the marvels of antiquity, the piece was the pride of both Pope Paul II and Lorenzo de Medici. It is called the Farnese Cup, after the Italian family that acquired it in the fifteenth century. Many eminent scholars, from Doctor Adolf Furtwängler to D. B. Thompson, have provided provenance (source or origin information) about the piece but cannot settle on a positive date for its execution. Estimates of the date range from 181 to 173 B.C. to the first century B.C. Its place of discovery is uncertain, but there are claims that it was found in the Villa Adriana, near Rome.

The Tazza Farnese is carved from sardonyx in shades of brown and pale ocher, all warm tones, with areas of translucent white material. The front shows a scene that symbolizes the agricultural bounty of the Nile River, with a sphinx in the foreground with a white head and dark body. There are seven symbolic figures: The river god Nile, Horus, Isis, two nymphs representing the seasons of flood and harvest, and two figures representing the winds. It has also been interpreted as the apotheosis of the first Ptolemy, with Ptolemy Philadelphus consecrating the festival of the harvest, instituted by Alexander the Great. The cameo has been discussed and interpreted, but scholars today still debate the true meaning of the scene.

The reverse of the piece has a raised head of Medusa with snakes and a scaly aegis. The powerful image of the Medusa was considered magical to the believer in spells and potions. It was said that the likeness of the Gorgon on a plate or cup would prevent death by poison. The vessel has a hole in the center, probably meant for the attachment of a stem.

## Roman Cameos

Alexander the Great's death in 323 B.C. left a leadership void that could not be filled. His empire collapsed, divided into dominions ruled by independent kings who had been generals under Alexander. Many of the countries, including Greece itself, were subsequently captured by Rome and incorporated into the Roman Empire. The result of such shifts in power was a massive infusion of Greek art, literature, and philosophy into Roman culture. The Romans may have been the conquerors, but they were enslaved by Greek sculpture, art, religion, and literature. They also acquired new tastes and lavished upon themselves the luxuries and wealth that came from their new possessions. During the first century B.C., Augustus became the first emperor of Rome, and the republic was forged into a mighty empire. It was at the time of the Caesars that cameo art was at its Hellenic best, with the summit of technique and development reached about A.D. 70. This is an important benchmark for cameo researchers because the Augustan era was later considered a perfect and classical period of the art, and the style was slavishly copied by the Renaissance carvers hundreds of years later.

A lapidary of note was Dioskourides, who carved a portrait of Caesar Augustus for the emperor to use as a personal seal. The Greek gem carvers were still expert in this field and carried on their trade while teaching the craft to the Romans; most of the Roman art of this time was in fact produced by Greek artists. An interesting note on dating cameos from this era is that Greek engravers continued to prefer mythological and heroic subjects for their cameos and intaglios rather than designs contemporary to the time. Greeks left their figures nude, but Romans prefered their figures draped.

Cameos also became fashion accessories and were used on helmets, breastplates, sword handles, and phalerae, fastenings for armor. According to Pliny the Elder in his *Natural History of the World,* A.D. 79, cameos were worn as insignia with ceremonial dress. Cameos were perfect accessories for the short loose cloaks called chlamys. The rich and noble collected cameos and filled cabinets with them for the pleasure of collecting. Some placed their pieces in temples as an offering to a deity. Anyone with the slightest pretension to taste and culture formed cabinets of gems. The rivalry that developed among cameo collectors resulted in an unprecedented rise in value.

Cameos were widely used in rings and other jewelry, as well as on crowns, cups, vases, dishes, and other ware. And the Romans believed the bigger the cameo the better, using them as decorations for the home as well as ornaments for vessels of utility.

The most important development for the cameo in this period was the use of portraits of nobles, generals, and heroes. For the most part, however, the designs and subjects of the cameos reflected the fashions of the day, with mainly pictorial motifs. Occasionally the name of the owner of the cameo was included in the carving, but usually the subject was a replica of a statue of a god or goddess (fig. 1-4) or hero (color plate 2). The most frequently recurring subjects on Roman gems depicting the deities were the figures of Jupiter, Minerva, and Juno, as well as some especially worshipped in Rome, for instance, the goddess Roma, who was adapted from the Greek goddess Athena, patroness of Athens. (The goddess Roma is distinguished from Minerva, another adaptation of Athena, in that Roma is seated on a throne and holds the orb and scepter of the world.) In fact, many of the Roman deities were based upon the Greek. The most famous trio of early Latin worship, Jupiter, Juno, and Minerva, were modeled after

**FIGURE 1-4.**

*A late Hellenistic cameo from the first century B.C., showing Psyche, defined by her butterfly wings, driving a biga drawn by two winged horses. The cameo is in its original gold ring mounting and measures about 1 by 1 inch (25.3 by 22.6 millimeters). (Content Family Collection of Ancient Cameos, Houlton, ME)*

---

the Greek gods Zeus, Hera, and Athena. Mars (Ares), Venus (Aphrodite), and Cupid (Eros) (figs. 1-5 and 1-6) appeared in a variety of representations. Other popular subjects, such as Bacchus (Dionysus) in his exploits, along with those of his followers, the sileni, wise but drunken old men (color plate 3), were interpreted by carvers in the traditional ways because they understood that the gods were metaphors for real human emotions and experiences. The figure of Hygiea with her serpent (fig. 1-6), the goddess of health and the daughter of Aesculapius, and numerous allegorical figures, such as Fortuna, Abundance and Indulgence, were also popular subjects. The Romans also liked to wear cameos carved with portraits of long dead, much admired Greeks, feeling a powerful affinity to that culture.

Many wealthy Romans had slaves of Greek origin who were skilled engravers and lapidaries. It was not at all unusual for a Roman master to act as patron for a business venture and set up a skilled slave or freeman in the jewelry trade. In his book *A Roman Book on Precious Stones,* Sydney Ball points to a Roman inscription found in the ruins at Malton, Yorkshire, England, that mentions a goldsmith's shop run by a slave: "Good luck to you slave in running this shop." Because the Romans had such an unquenchable passion for cameos, hundreds of Greek artists were attracted to the Roman capital to work. These artisans produced literally thousands of seals and cameos in an array of materials including glass (fig. 1-7). All workmen did not have equal skills, however, as every degree of craftsman-

**FIGURE 1-5.**

*Eros, wearing a chlamys, plays a lyre. Dated as Roman, first century* A.D., *the onyx cameo exhibits a raised line border. It measures about ½ by ½ by ½ inch (15 by 13.8 by 2.2 millimeters). (Content Family Collection of Ancient Cameos, Houlton, ME)*

**FIGURE 1-6.**

*Two of three cameos found together on a necklace in Syria dating from the third century* A.D. Left: *An onyx cameo of Eros teasing a goose; note the hairline crack across the gem.* Right: *An onyx cameo with Hygiea facing left. She wears a himation, her hair is tied back, and she holds a serpent in one hand and food in the other. (Content Family Collection of Ancient Cameos, Houlton, ME)*

---

ship from master to mediocre can be seen in the gems of this era. During the Roman period, gem engravers were called *gemmarii* and cut either cameos or seal intaglios. The cutters of only cameos were called *caelatores,* while those who specialized in intaglios were known as *cavatores* or *signarii.* These specialists were organized in loosely knit groups that became the earliest of the artisan guilds.

**Figure 1-7.**

*Two white-on-blue-glass cameos that have been dated from* A.D. *400.* Left: *A maenad defends herself from a satyr; a popular Roman theme.* Right: *Eros walking, with the attributes of Heracles—in his right hand a quiver and over his shoulder the lion's skin and a club. Repaired. (The Sommerville collection, University of Pennsylvania)*

Julius Caesar (100 to 44 B.C.) was a great lover of seals and cameos and owned a signet seal of a standing figure of Venus Victorix with a palm in her hand. This seal flattered his pride of ancestry, as he was fond of claiming that he was descended directly from the gods Venus and Aeneas. During his reign Caesar collected at least six cabinets of engraved gems and cameos for dedication to the temple of Venus.

The Gemma Augustea (Coronation of Augustus) and the Agathe de la Sainte-Chapelle (Grand Camée de France), two of the most celebrated cameos ever carved, were completed in the Augustan era.

The celebrated Gemma Augustea is a two-layer elliptical black and white sardonyx cameo, approximately 8 by 9 inches (color plate 4). It is the crowning gem in the impressive collection of cameos in the Kunsthistorische Museum in Vienna. While there is no way of knowing the value of the Gemma Augustea when it was made, it is known that the Emperor Rudolf II (1552 to 1612) of the Holy Roman Empire paid 12,000 gold ducats for it in the sixteenth century, the equivalent of more than half a million dollars today. The historical and artistic value of the cameo is incalculable, as it is without question one of the finest jewels of the ancient glyptic art.

The cameo has a black background with relief figures in white in a stratum so translucent and thin in some areas that a bluish shadow and light appears. Shown are members of the Julian and Claudian ruling families of Rome. As was typical of cameos of that time, the princes are represented as gods, and the figures of the huddled masses symbolize cities and their people. There are two distinct levels in the cameo. In the top level, Augustus is shown seated on a subsellium—the symbol of his tribunician power—near the goddess Roma. Under his feet are weapons from victories. On the ground near Augustus is the eagle, a symbol of the god Jupiter and the city of Rome. Above Augustus in the sky is the zodiacal symbol for Capricorn, the constellation under which he was born. Augustus, half nude, leans on a scepter while behind him the symbol of the world, Orbis Romanus, holds a crown above his head. Seated behind him are the Ocean and Terra Mater, indicating that the Ocean (the sea) is pacified and Terra Mater (the earth) is producing her fruits again. In front of Augustus and Roma, Tiberius, crowned with laurel and with a scepter in his hand, steps down from his chariot. Victory stands beside him in his chariot. Germanicus, nephew of Tiberius, stands next to the chariot. Tiberius has won a war with the Pannonians, and in the lower panel Roman soldiers build a trophy while Pannonian prisoners beg for mercy, sit in despair, or are pulled into the victory square by Roman soldiers.

Large cameos such as the Gemma Augustea were too big to be used for dress ornaments and were probably meant as wall or cabinet decorations. Many of the cameo-encrusted cups and dishes in museum collections were strictly for decoration despite their functional appearance.

The cameo called the Sainte-Chapelle, or Grand Camée de France (fig. 1-8) is part of the collection in the Bibliothèque Nationale in Paris. It is a five-layer sardonyx oval stone, 13 by 9 inches, engraved in Rome by Dioskourides in the first century A.D. This piece has been described in the ancient inventories as "Le Grand Camahieu," "Le Grand Camée de la Ste.-Chapelle," and "Agathe de Tibere." During the Middle Ages, the cameo was venerated as a holy relic, and it was understood to represent the triumph of Joseph in Egypt. Not until the seventeenth century was it decided that the proper interpretation of the subject was the apotheosis of Augustus.

The cameo has a fascinating history. It was taken to Constantinople with other imperial treasures by Emperor Constantine I and stayed there until 1244 when Baldwin II, a Latin monarch and the last Frankish emperor of Constantinople, tried to pawn it as part of his treasures along with the Crown of Thorns, swaddling clothes of the infant Jesus, and other religious relics to Louis IX, king of France, for large sums of money to defend his throne.

The design of the cameo is allegorical and therefore has many interpretations. It consists of three scenes. In the upper portion is the apotheosis of Augustus. In the middle section, Tiberius is seen under the

FIGURE 1-8.

*The Grand Camée de France was venerated as a holy relic in the Middle Ages. (From Hodder M. Westropp,* Handbook of Archaeology *[London: Bell and Daldy, 1867], p. 273)*

figure of Jupiter, and Livia, his mother, under the figure of Ceres. The pair are receiving the young Germanicus upon his victorious return to Rome in A.D. 17. Behind Germanicus, Agrippina, his wife, helps him take off his helmet, and his young son Caligula stands behind him. A young man carrying a trophy and identified as Tiberius Drusus is seen behind Germanicus. The bottom portion of the cameo displays figures and warriors dressed in costumes of eastern and western people, personifying conquered nations.

Seals remained of great importance to the Romans. In fact, the fate of a signet ring after its owner's death was of prime importance to his family, heirs, and friends. Pliny writes of heirs who solicitously crowded the bed of a dying man to be able to oversee personally the fate of his signet. According to Pliny, some Romans left implicit instructions that their signet seals be enclosed in their funeral urns. However, if the dying gave no instructions for the future of the signet, or if the end was too sudden to make plans, the signet seal was unceremoniously snatched from the finger of the owner at the moment of death.

The period of the principate of Augustus (30 B.C. to A.D. 14) was the golden age of ancient glyptics, and it was marked by an incredible profusion of intaglios and cameos. A new method of carving on small agates, called nicolo, was developed and widely used in Roman glyptics. Nicolo is the term given to those cameos carved in onyx with a thin layer of faint bluish-white material over a thick layer of black; the design is cut in the upper translucent bluish-white layer so that the black background shows through, giving the design a bluish tinge.

Pliny recounted that in the first century A.D., the Romans adopted the fashion of wearing gems engraved with an image of one of the Egyptian deities, because the sects of worship of Isis and Serapis were flourishing. Also, gems carved with lucky horoscopes were used in the Roman Empire, where superstitions of all types were prevalent. Talismans and amulets that purported to bring good luck or have some other magic powers were highly valued. Eroticism was a part of a love cult and a type of sympathetic magic; the amulet both identified the enchantment and titillated the owner (fig. 1-9). The power of the talisman depended partly on the material from which it was cut, partly on its color, and partly on the subject of the carving itself.

FIGURE 1-9.

*A fragment of a Roman erotic cameo in sardonyx from the late second century to early third century A.D. Scenes of lovemaking are found on both Greek and Roman gems. (Content Family Collection of Ancient Cameos, Houlton, ME)*

The grylli, a fanciful combination of joined heads carved on a cameo, were interesting subjects in this era. A nineteenth-century interpretation of classic Roman grylli invokes the spirits of two deities for good fortune (color plate 5). The grylli cameos were often combinations of several beasts, humans and beasts, or humans, insects, and beasts. The subjects were always tied to the mystical and mythological.

Wearing gold coins with the portraits of earlier emperors became fashionable among the populace in the third and fourth centuries A.D.; it was a custom imitated much later in other societies. The passion for coins did not stifle the engraving of gemstones, but it did cause a decline in skills and design. By the time of Constantine in the fourth century A.D., the craft of gem engraving and cameo carving had, like most of the other arts, sunk to a very low ebb. As the Roman Empire collapsed, the cameo, like the world itself, entered the Dark Ages.

## The Dark and Middle Ages

As the barbarians conquered Rome and ancient civilizations crumbled, only a few groups remained to keep centuries-old arts alive. In the beginning the Christians lived simply and with contempt for riches. To them, art, including engraved gems and cameos, signified idolatry, especially when repeated in sculpture. The early Christian church barely tolerated art or dismissed it altogether. They adopted symbols such as the dove, fish, anchor, ark of Noah, and boat of Saint Peter as their own and used them on seals and cameos so that the Christians might identify each other more easily. For the most part, however, much of the recorded wisdom and treasures of past cultures were buried, lost, or destroyed, including sculptures, gem engravings, and cameos. Because Christians did not bury valuables with the dead, much of their early art was not preserved. So, with knowledge lost and training and education in the arts denied, the dark age of cameo carving was set into motion.

What survived from the era was largely due to monks, who kept the flickering light of art alive with their illuminated manuscripts. A number of the manuscripts had margins decorated with crude, often stiffly drawn or carved cameos in bone or ivory. Many other liturgical objects were decorated in a like manner, but the designs on the cameos were often crude to the point of being ludicrous (fig. 1-10). Many of the carved stones are simply portraits of unknown persons, unfathomable in meaning and poorly crafted. But, even though there was little regard for the artistic value

**FIGURE 1-10.**

*A crude cameo from the Middle Ages. (The Sommerville collection, University of Pennsylvania)*

of antique cameos, they were highly sought as talismans or amulets by the sick and weary. Both the subject matter and the material were believed to have prophylactic attributes.

It became custom from the fourth century on to commemorate important people and historical or religious events by carving diptychs and triptychs (two- and three-panel tablets) in ivory or wood. Scenes from the panels were later much copied by Renaissance artisans in cameos. An artisan might take one figure, group, or pose from a scene for the entire design of a cameo. This is most often observed in cameos with religious designs. The surviving fragments of basso-relievo from book covers, diptychs, and triptychs present vital information on the remnants of cameo carving in the fourth to sixth centuries.

A new type of cameo carving flourished in the fifth century, classified as the motto cameo. The carvings are short sentences or mottos cut in a circle on a stone, sometimes closing a symbol. These cameos were designed as small gifts (figs. 1-11 and 1-12). The mottos were appropriate to the occasion for which they were supposedly engraved: "Long Life to Thee," "Mayest Thou Live Many Years," "Prosperity." A frequently used motto, probably given upon the departure of a friend or lover, was "Remember me, thy pretty sweetheart"; it was generally accompanied by a picture of a hand pinching an ear, the belief at that time being that the ear was the seat of memory (fig. 1-13). This hand tweaking an ear had been commonly used in Roman art and literature, meaning to request attention. Similar stones in the British Museum are reminders to a loved one with messages: "Remember me, your dear sweetheart," and "Remember me, your love, wherever you are."

As monasteries began to flourish, the Roman Catholic church

FIGURE 1-11.

*A massive Roman finger ring with a white-on-brown onyx cameo depicting a clasped right hand. The wrist on the left is encircled with a bracelet and identified as female. At the top of the cameo is a garland; below is the legend* OMONOIA *("Concord"). The cameo measures about ½ by ½ inch (15.3 by 14.2 millimeters). (Content Family Collection of Ancient Cameos, Houlton, ME)*

FIGURE 1-12.

*This solid gold mounting, set with the cameo shown in figure 1-11, is in excellent condition. The mounting measures about 1⅛ inches (30 millimeters) wide. (Content Family Collection of Ancient Cameos, Houlton, ME)*

---

became the recipient of wealth and riches that the monks gladly donated in their pursuit of an ascetic life-style. Although the monks had creative abilities, it is generally believed that an illuminated manuscript with superior artistic qualities, highly decorated with carved gems, is from the hand of an artist who lived much later, perhaps as late as the Renaissance, and in a less restricted environment, where he could draw his illustrations from life. Scholars point out that the creative abilities of monks in the fourth to sixth centuries were stifled because they had no access to live models and models from antiquity had been destroyed. As the monks had

FIGURE 1-13.

*An exquisite example of a sentimental ancient onyx cameo, about 1 by ⅞ inch (25 by 22 millimeters), depicting a hand pinching an earlobe with the legend* MNHMONEYE MOY *("Remember me"). This type of cameo would have been given as a special keepsake to a lover, friend, or family member. It is set in an eighteenth-century swivel gold ring mounting. (Content Family Collection of Ancient Cameos, Houlton, ME)*

no women around them, they had to draw or carve from memory, dreams, and their conceptualization of the ideal.

Most interesting is that the old pagan motifs were recycled and renamed to fit the new religion. Where there was no subject model, one was developed. The first simple pagan symbols borrowed by the Christians were the palms of Egypt, and the vines and wreaths of Greco-Roman myths. The Hellenistic Helio or Orpheus became the youthful, beardless Christ. As cameos with Latin subjects now were interpreted as Christian legends and replaced other figures, Christianity was increasingly reflected in the arts. Monks renamed the seals and cameos to conform to their own iconography and interpretation of the subject. Every veiled Roman woman's profile became the Madonna (color plate 6); Jupiter became Saint John, even with Jupiter's symbol of the eagle by his side (color plate 7); the enthroned emperor of Byzantium carved seated was renamed as Christ the philosopher among his followers; Medusa became Saint Veronica; and Leda and the swan symbolized the Annunciation (color plate 8). All the old myths gave way to the new iconography, as Christian teachings were expressed by the ancient Greek and Roman cameos carved hundreds of years before the birth of Christ.

Meanwhile, in the Middle East, the craft of stone carving was cultivated strictly for the production of seals. The art of engraving could not progress, however, as the development of Islam and its religious code forbade the representation of all living things, human or animal. The result is that engraved stones of Muslim cultures have only inscriptions on them, such as the name of the owner, and frequently prayers or passages from the Koran.

In the eleventh century, classical cameos were sought after to be used in pendants; they were particularly popular in Ottonian Germany. Apparently a vast collection of cameos and engraved gems was excavated around the city of Mainz, but they were subsequently lost during the years of World War II, or at least they have not resurfaced for public display.

Some cameos from the eleventh century do exist for study, displayed in major European museums, including those in Vienna, Pforzheim, Munich, and Berlin. Most of these pieces display images that conveyed messages to a largely illiterate population. It can only be surmised that one of the uses of the cameo during this period was as a type of religious icon, to invoke in the owner a sense of divinity, to attract and hold the attention of the believer, and to guide and direct prayers.

Throughout the Middle Ages, royal and religious patronage kept artisans working and producing. In the early twelfth century, some brooches were being worn that were set with bas-relief carved stones; and by the early thirteenth century, an interest in the classical past was developing, along with a new demand for cameos and engraved gems. The carved gems were used mostly as decorations for reliquaries and secular objects. A few extraordinary cameos were produced for the Burgundian court in antique style, but most jewelry from this period is lost, having been melted down by its later owners. Information about jewelry of the thirteenth century is mostly gleaned from royal inventories that survived with household account documents. It is recorded that, in the middle of the thirteenth century, a law was passed in France forbidding commoners to wear any precious stones, pearls, gold, silver, cameos, or engraved gems. Jewelry became a privilege reserved for a very few people, and it was not until the end of the century, when the sumptuary laws were lifted, that jewelry slowly made its way back into the life of the ordinary citizen.

## The Renaissance and Later

From the Renaissance forward, there was a new interest in classicism, and the rebirth of learning began with the study of ancient Greco-Roman arts and literature. As the influence of the classics spread, there was a transformation in art, literature, science, and religion. This revival of the arts coincided precisely with the development of a prosperous middle class of citizenry, as well as the raising of women to socially prominent figures, who were anxious to give their patronage to jewelers. Jewelry became a part of costume, depending upon the whims of fashion, which itself reflected the tastes of a luxury-loving society.

The sculptors of the fourteenth century drew their inspiration from any antique art they could find around them while at the same time giving rein to their own creative imagination. The rebirth of the arts had a strong effect on cameos. The decline of craftsmanship of cameos and intaglios suddenly reversed, with better tools and technical skills, as well as the many, varied influences upon the artisans of the fourteenth century. Many were skillful in several different fields: as builders, architects, painters, engravers, and sculptors. Artists began to learn about perspective and light and produced marvels of art never before achieved.

Florence, Italy, was the great center of artistic activity, and the great sculptors and painters, carvers and famous architects of the time were either natives of that city-state or resided there. Florence reached the height of prominence as an art center during the reign of Lorenzo de Medici—Lorenzo the Magnificent—an ardent patron of the arts. It has been written that while the Medici family were great patrons of the Italian Renaissance, they were not particularly collectors of gemstones, amassing instead other types of art: tableware for special occasions and precious religious objects. It is well known, however, that Lorenzo was a true collector and connoisseur of stone carvings. He had many cameos carved for him and had them engraved "Lor Med," as one might stamp one's name on personal property. His name is on a finely carved cameo of Noah and the ark, engraved in tiny letters on the ark. The jewel is now in the British Museum, acquired in 1890. Lorenzo was also the one-time owner of the celebrated Tazza Farnese cameo.

The subject matter and designs of cameos of this era were mainly portraits and Greco-Roman myths (color plates 9, 10, 11, and 12). The shift in design during the Renaissance, however, was not to a new style but to a revival of the ancient and classical. Craftsmen copied the old cameo designs, as well as creating new ones using the old stories and legends. Many cameo portraits were executed in profile; the twelve Caesars, as well as Alexander the Great, were some favorite subjects.

One of the earliest patrons of cameo art of the period was Cardinal Pietro Barbo, later Pope Paul II. This church official acquired an outstanding collection, which he is said to have kept in silver boxes. The story goes that the pope was killed by his extravagant display of his carved gems because he wore so many cool stones on his fingers mounted as ringstones that he caught a chill that led to his death. Subsequently, the pope's cameo collection was acquired by Lorenzo de Medici, to add to his collection begun by his grandfather, Cosimo the Elder (1389 to 1464). Much of the great three-thousand-piece collection is now divided among museums in Naples, Florence, and Paris. Although the level of the craftsmanship of cameos of the Renaissance is the finest, there are no cameos with the extraordinary size or reputation of the Gemma Augustea or Sainte-Chappelle. Cameo historian Cyril Davenport has suggested that the sizes of the cameos, their designs, and their framing serve as a means of dating them. Davenport noted too that both raised borders around the design of the same level as the carving and color are signs of fifteenth-century work but conceded that these are found occasionally in ancient work. He also suggested that cameos with signatures cut in relief are indicators of craftsmen contemporary work; it seems they were signed when the design was carved, whereas a signature cut in intaglio may have been a subsequent addition.

There was tendency during the fifteenth century for carvers to use the dark layers of onyx as portraits of the heads of blacks, for which various explanations have been offered from the nobles' fondness for having black servants to historian C. W. King's belief that the profiles were intended to commemorate the black concubine of Clement VII. It has been written that Clement's mistress caused the Florentine school of cameo art to produce dozens of gems showing Ethiopian women.

During this time cameos were not carved as amulets or charms, as they had been in Roman days. While the ancient myths, gods, and goddesses were copied from the classics, they were regarded as art objects and considered a sign of high intellect. Collectors were swept up in a mania of collecting scenes of Eros and Psyche (color plate 13) and Leda and the swan (fig. 1-14). The cameos that were worn on clothing were usually mounted in what is known today as brooches. The brooches were worn on the shoulder, in front of the garment, or, in the case of a small cameo, as a ring or necklace. Although it looks like a brooch to the twentieth-century collector, another cameo jewel was the hat badge, called *enseigne.* The badges were worn on the front of a hat or cap and were fashionable for men from 1520. They can be seen in numerous paintings from the Renaissance. The best examples of hat badges came from Italy, and two of the most popular subjects were religion and myth. Model examples of hat badges can be found in the British Museum, one depicting the judgment of Paris, another the conversion of Saint Paul. The badges originated from the medieval pilgrim insignia, sold at religious shrines. Men with classical education preferred antique cameos mounted in a contemporary setting,

**FIGURE 1-14.**

*One of the most often repeated images in carvings is Leda and the swan. This gray and white sardonyx cameo, set in a ring, has been dated to the neoclassic period (A.D. 1700 to 1850). (The Sommerville collection, The University Museum, University of Pennsylvania [neg. # 139550])*

---

with a fitted loop for attachment to a beret. The religious subjects were generally thought to indicate the devotion of the wearer to his patron saint.

Because there was a scarcity of acceptable carving material in the sixteenth century, some carvers began using beads acquired from India, cutting them in half and carving upon the domed surface. When that supply was depleted, they substituted genuine ancient cameos, sometimes working on the backs of the gems, but not infrequently eliminating the previous design altogether and carving their own subjects onto the material. When the great supply of German agate ran out, other materials were investigated. The search led to the use of shells, which had several strata in different colors. Shell became popular as a medium for carving, especially for mass-produced objects, because it could be carved quickly with hand tools. The majority of shell cameos were carved in France and Italy; researchers note the first mention of shell cameo work in literature is by Fra Borghigiani of Florence. Shell was exclusively used for cameos; intaglios were rarely, if ever, attempted. Cameos from this period will be found in shell with gray backgrounds. If genuinely antique, the cameo will likely be highly abraded because of the softness of the shell and its susceptibility to deterioration.

Elizabeth I of England is credited with introducing the custom of giving cameo brooches or pendants as gifts to her loyal subjects for favors or as payment for a particular service. The most widely acclaimed example of a cameo with Elizabeth's portrait is the Armada jewel, designed by Nicholas Hilliard and on display in the Victoria and Albert Museum in London. The jewel is a two-sided pendant with a cameo of Elizabeth in profile on one side. The cameo is mounted on a background of fine blue lapis lazuli; the mounting is decorated with enamel and set with diamonds and rubies. The back of the jewel is engraved and enameled with the design of an ark caught in a storm.

If cameos had a romantic period, it was during the eighteenth century, when schools of engraving were founded and the passion for engraved gems had reached epidemic proportion. Every man yearned for a crest of his own as a mark of prestige and culture.

From all over the world, goldsmiths and artists flocked to Rome and Florence to take instruction in the art of gem carving. In the carving schools, would-be artists/carvers studied classical subjects and copied them in detail. Later, some of the most artistic and talented carvers were dismayed to find their school exercises turning up on the market as "genuine antiques." But, of course, a few seized the opportunity to commit fraud themselves, by signing the name of a real or fictitious Greek carver to their works. This was apparently a lucrative endeavor, as cameo collecting was enjoying another wave of popularity. Mass production was needed to meet the demand, and cameos were being rushed to the market without the same careful carving displayed in earlier pieces. Lack of quality did not seem to matter, however, since new, less sophisticated collectors had little sense of aesthetics or knowledge to differentiate between a classical cameo or a contemporary fabrication.

Though gems were poorly carved, they remained a cultural status symbol. A letter by a women to her lover, quoted in Anne Garside's *Jewelry, Ancient to Modern* supports this conclusion. In the letter she mentions that a friend visiting her "says seals are much in fashion, and by showing me some she has set me alonging for some too." Farther along, the lady complains of her friend's excesses: "She wears twenty strung on a ribbon, like nuts boys play with, and I do not hear of anything else."

The period also saw a rise in travel, and leading the list of tourist sights were the newly uncovered ruins of Pompeii and Herculaneum. Tourists, then as now, were of a single mind when it came to taking home souvenirs to remind them of their holiday. Shell and lava cameos, inexpensive, easy to pack and carry, and indigenous to the area, became the souvenirs of choice. Shell had become a quick and easily obtained medium for portraiture (color plate 14); abundant lava from Mount Vesuvius in colors ranging from olive green to cream was commonly carved into portraits of gods and philosophers (fig. 1-15).

Fashionable women particularly enjoyed wearing such shell and lava cameos because their inclusion in a costume signified a traveled person of cultivated taste. Many interesting and amusing shell cameos showed scenes of cupids, after the brilliantly colored frescoes in the magnificent House of Vettii in Pompeii (fig. 1-16 and color plates 15 and 16). The cupids in the original works are shown performing the daily chores of various occupations, from goldsmith to flower vendor in Pompeii of A.D. 79. The excavations aroused interest in archaeological jewelry and revived the use of classical motifs, as well as creating a vogue for the antique.

After Napoléon's 1796 Italian campaign, the clamor for cameos grew even stronger. The French desire was fanned by Napoléon's own

**FIGURE 1-15.**

*Lava has been a popular medium for carving cameos since the early eighteenth century. Shown are late-nineteenth-century lava cameos.*

delight in cameos; and he brought back many pieces from Italy for himself. Impressed with the art form, Napoléon promoted cameo carving in France and founded a school in Paris especially to teach cameo carving. Cameos were even part of his wedding regalia, and one of the most beautiful items from this period is the coronation crown, Couronne de Sacre, made for Napoléon's coronation in 1804 and worn at his wedding to Marie-Louise in 1810. The gold-plated silver crown consists of a circlet of eight laurel leaves that rise from four crossing arches, surmounted by an orb and cross finial. The crown is decorated with dozens of fine antique Roman stone cameos.

Along with the poor and the mediocre, the eighteenth century can boast some of the most celebrated carvers: Lorenz Natter, Jacques Guay, Giovanni Pichler, and others. A problem in identifying the many cameos of this period is that some of the well-known carvers copied works from the classical Greek, and some totally unknown carvers signed the names of Pichler, Natter, and others to their work. Trying to authenticate a piece can result in hopeless confusion, especially since some of the work is so

FIGURE 1-16.

*Two late-twentieth-century shell cameos, carved in Torre del Greco, Italy, shown with their inspiration, the Pompeiian fresco in the House of Vettii.*

technically correct that no anomaly or innovation can be found to mark the cameo as eighteenth century instead of ancient. The origins of some gems have been debated for years by scholars, without success.

Giovanni Pichler, one of the best-known carvers, had the opportunity to instruct Madame de Pompadour in the art of cameo carving, and he enjoyed her patronage for many years. Catherine the Great of Russia gave employment to the English brothers William and Charles Brown for over ten years. They filled her cabinets with finely carved cameos; at one time more than four hundred stones, half of their total oeuvre, were displayed in the Hermitage. The subject matter of Catherine's collection ranged from myth to history of allegorical subjects and animals. One of the most fascinating was a series depicting the Russian victory over Turkey in the war of 1787 to 1791, complete with symbols typical of the time. Catherine was so caught up in the cameo-collecting mania that she added to her cabinets by buying a complete set of the glass replicas of cameos and intaglios from James Tassie, a Scots manufacturer of molded-glass imitation cameos in London. The empress commissioned Tassie to make her as complete a collection as possible of ancient and modern carvings in paste reproductions and in white enamel. He complied with over twenty thousand pieces.

Another passionate collector of the art was Empress Joséphine, Napoléon's first wife. She went so far as to break up some of the old royal jewelry taken from the kings of France so that she could make up complete jewelry parures, like the one pictured in color plate 17.

Empress Maria Fyodorovna, widow of Paul I of Russia, gained a reputation as a good and clever cameo carver in the late eighteenth and early nineteenth centuries. She was a member of several European art academies and learned the art of cameo carving firsthand from a German named F. Carol Lebrecht, head of the Russian mint and himself a noted engraver in the late eighteenth century. Lebrecht founded a school in Russia where he trained many in the glyptic arts.

In the nineteenth century, the popularity of shell cameos was revived and its daily wear became fashionable once more. Carvers made the easy transition from hard stones to the softer shell materials but were apathetic in any event because mass production and assembly-line techniques had turned the once skilled and eager craftsman into an anonymous pair of hands. Cameo carvers of the early nineteenth century were, for the most past, without the technical virtuosity of their predecessors.

In this indifferent atmosphere of the nineteenth century, the anonymous women's profiles became the popular subject for cameos. They could be replicated quickly, and if the original model ever had a known identity, it was lost in the repetition and reinterpretations from workman to workman.

In the first half of the nineteenth century, the glyptic arts were snuffed out quite successfully when fakes and forgeries took hold in the world art markets and created such scandals that collectors lost all interest, especially when they began to realize how easily they could be fooled. The cameo market went into a decline, and by the turn of the twentieth century, cameos were no longer a revered art form; the general public lost interest. Cameos for use in jewelry were still being made on a limited basis, and those in the costume-jewelry business struggled to promote them along with semiprecious stones as fashion items. In the early 1900s, two celebrated European theater actresses, Sarah Bernhardt of Paris and Eleonora Duse of Italy, were favorite models for the cameo carvers, as they were the role models for the liberated women of the day. Today many cameos can still be found with the portraits of Bernhardt and Duse dressed in the costume of one of their stage personas.

World War II interrupted the cameo-carving industry, already on the decline. After the war the demand resumed slightly but never enjoyed a full-scale revival.

Collectors and scholars believe that the decade of the 1990s will see a true cameo revival that extends into the twenty-first century. The reasons are many and varied: collectors are attracted to this "rediscovered" and unique art form; investors believe the nineteenth-century gems will rise sharply in value; the artisan capable of hand carving is beginning to recognize an opportunity to supply creative work to a demanding market; a new generation of consumers with money and education are discovering this art form; and the mass-produced machine-made cameos have themselves generated new interest, creating a collectibles market.

Overall, cameos have a long and noble past. They have been—and can be again—one of humanity's greatest and most treasured art achievements.

*Chapter*

# 2

# THE SUBJECTS: MYTHS, LEGENDS, HISTORICAL EVENTS, AND STORIES BEHIND THE JEWELS

There is power in the carved image as much as in the spoken word, and that power was well understood by the ancient Greeks and Romans. Carved gems reveal the social structure and philosophy of their own particular time. For instance, the biblical story of Adam and Eve in the Garden of Eden would not be found on an ancient Greek or Roman cameo, but during the Byzantine era, Christian iconography was the norm. The articulation, if it is an original work of the period, is from an artist's own perspective, what he sees and believes as truth.

Ancient people did not have the science that we take for granted today to explain the events of the physical world. Early theories of the universe contended that the entire universe revolved around the earth. The ancient Greeks were convinced that the earth was a flat disc resting upon an infinite ocean of air and stars were silvery nails driven into the vast outer sphere. Anaximander, born in 500 B.C., was a Greek philosopher who taught the sun was a giant chariot wheel in the sky with a hollow rim of fire shooting out flames from a hole in the rim. Early Mesopotamians believed the earth rested upon the back of a giant tortoise; as the animal slowly moved forward, the sun and moon rose and set around it. With all of these fantastic explanations of scientific phenomena, it is easy to understand why myths and legends were so powerful in ancient civilizations and why early cameos depicted subjects that connected traditions, legends, religion, and myth. So many cameos were carved with myths that they may be classified as stone literature. Some were like tablets of faith that early owners wore close to their hearts. Cameos cut as icons of religion brought comfort to their owners and served as invocations to the gods.

A myth is one way of explaining the world and the things in it. Beliefs vary according to way of life; for example, to nomadic herdsmen, a strong and powerful sky god was the best protector, while to a peaceful agrarian society, the earth mother was all important. Myths frequently result in a hierarchy of gods that parallels the society of believers. Sometimes a hero will come forward, rationalize the myths, and give them formal shape, as Homer did for the Greeks.

There is a thin line between a myth and a folktale. A myth attempts to organize the origins of phenomena and explore human qualities such as courage and strength; it explains events and conveys beliefs. Folktales revolve around humans and their adventures and at the same time instruct and entertain.

Ancient Greek myths were of three types: those concerning the Olympian gods; those explaining natural events; and those reporting deeds and heroic adventures. Against this background of mystical imagery, the subjects and designs of cameos developed in the fourth century B.C. By learning and understanding some of the myths, the collector achieves a much greater enjoyment of the cameo art form, while jewelry historians and jewelry appraisers will, in a practical way, be better equipped to handle cameo circa dating.

## Myths and Legends

Foremost in the Greek pantheon were the twelve Olympians, the elite among the gods and goddesses. They were believed to dwell with Zeus, their lord, and for some, their father, on Mount Olympus. The list of the twelve Olympians was modified from time to time, but the most frequently named were: Zeus, Poseidon, Hera, Demeter, Apollo, Artemis, Aphrodite, Ares, Hephaestus, Hermes, Athena, and Dionysus.

The Greek myths were modified by cults from other lands, and the most important of the Greek gods were assimilated by the Romans. The chief Roman gods are equated with the twelve Olympians. In Western Europe from the time of the Renaissance until today, the Greek gods have been referred to by the names of their Roman counterparts. The accompanying table and list indicates the Greek and Roman names of the deities, as well as their dominions and the attributes associated with them.

*Greek and Roman Divinities*

| GREEK NAME | ROMAN NAME | DOMINION | ATTRIBUTE |
|---|---|---|---|
| *Aphrodite* | *Venus* | *Goddess of love and beauty* | *Swan, dove, myrtle tree* |
| *Apollo* | *Apollo* | *God of music, poetry, purity, and light* | *Golden lyre* |
| *Ares* | *Mars* | *God of war* | *Vulture, dog, golden plumed helmet, sword* |
| *Artemis* | *Diana* | *Goddess of hunting and childbirth* | *Bow and arrows, deer, cypress tree* |
| *Asclepius* | *Aesculapius* | *God of medicine* | *Staff, snakes* |
| *Athena* | *Minerva* | *Goddess of crafts, war, and wisdom* | *Owl, olive tree* |
| *Cronus* | *Saturn* | *In Greek mythology, ruler of the Titans and father of Zeus; in Roman mythology, also god of agriculture* | *A radiant head, scythe* |
| *Demeter* | *Ceres* | *Goddess of the harvest and fertility* | *A stalk of wheat or grain* |
| *Dionysus* | *Bacchus* | *God of wine and fertility* | *Grapes and wreaths of grape leaves* |
| *Eros* | *Cupid* | *God of love* | *Bow and arrows* |
| *Gaea* | *Tellus* | *Symbol of the earth, and mother and wife of Uranus* | |
| *Hades* | *Pluto* | *God of the underworld* | *Cerberus, the three-headed dog* |
| *Hephaestus* | *Vulcan* | *Blacksmith for the gods, and god of fire and metalworking* | *Anvil and forge* |
| *Hera* | *Juno* | *Protector of marriage and women; in Greek mythology, sister and wife of Zeus; in Roman mythology, wife of Jupiter* | *Cow, peacock, pomegranate* |
| *Hermes* | *Mercury* | *Messenger for the gods; god of commerce and science; protector of travelers, thieves, and vagabonds* | *Winged sandals, wings on his hat* |
| *Hestia* | *Vesta* | *Goddess of the hearth* | *A hearth with a never-ending flame* |
| *Hypnus* | *Somnus* | *God of sleep* | |
| *Poseidon* | *Neptune* | *God of the sea; in Greek mythology, also god of earthquakes and horses* | *The trident* |

| Greek Name | Roman Name | Dominion | Attribute |
|---|---|---|---|
| *Rhea* | *Ops* | *Wife and sister of Cronus, mother of Zeus* | |
| *Uranus* | *Uranus* | *Son and husband of Gaea and father of the Titans* | |
| *Zeus* | *Jupiter* | *King of the gods, god of weather and fertility* | *Thunderbolt, eagle* |

Other mythological figures frequently found on cameos are depicted in a variety of designs, including:

*Abundance:* Portrayed with wheat and inverted cornucopia.

*Achilles:* A Greek warrior and the hero of the Trojan War; the central figure in Homer's *Iliad.*

*Adonis:* Wears a hunter's costume; usually with Venus; also with a dead boar and dog.

*Aeolus:* Guides Bacchus to Ariadne; he is a bearded, winged figure.

*April:* A youth dancing in front of a statue of Venus.

*Atalanta:* A Greek maiden, suckled by a bear, who grew up in the forests and was skilled in hunting and athletics. She declined to marry any man who would not risk his life in a footrace with her: if he won, she would be his wife; if he lost, he would die. She eventually married Hippomenes, who defeated her by throwing golden apples in her path, which she stopped to retrieve.

*Atlas:* Bearded nude figure, seated on a mountain.

*Bacchantes:* The female followers of Dionysus (Bacchus), also known as maenads, frequently shown dancing with flowers, grapes, and grape leaves in their hair.

*Castor and Pollux:* The dioscuri; they wear oval helmets and are mounted on horses. A star is their symbol. They are children from the union of Leda and the swan (Zeus).

*Charon:* The boatman who ferried the souls of the dead across the river Styx to the underworld. Often depicted with Cerberus, the three-headed watchdog of the underworld; sometimes with Hermes, who guided the souls of the dead to Charon.

*Concordia:* Holds an olive branch and stands between two standards.

*Deianira:* The bride of Heracles. She was responsible for killing the mortal part of him.

*Endymion:* Seen asleep in the arms of Morpheus; sometimes portrayed with Diana and preceded by Cupid holding a torch.

*Eos:* Sometimes shown as a charioteer with two horses, she is especially known as the lover and seizer of handsome youths and as the goddess of the dawn.

*Erotes:* Cupid or cherubs with wings.

*Europa:* A Phoenician princess carried by Zeus, in the form of a bull, to Crete, where he made her a queen.

*Fates:* Called Moirae by the Greeks, Parcae by the Romans. The Fates are represented as three women who dispense each person's lot in life, assigned at birth: Clotho spins the thread of life, Lachesis determines its length, Atropos cuts the thread.

*Hebe:* Goddess of youth, sometimes appearing as cup bearer to the gods or feeding the eagle of Zeus.

*Heracles:* The greatest of all slayers of giants and monsters; a never-resting performer of seemingly impossible tasks. Most familiar as a muscular man clad in a lion skin with a club and a bow and quiver.

*Flora:* The Roman goddess of flowers, resembling the Greek goddess Chloris.

*Fortune:* Depicted with the sun and crescent; two cornucopias; one foot on prow of a ship; as Fortuna Manes, holding the bridle of a racing horse.

*Graces:* Three daughters of Zeus, personifying beauty, charm, and grace—Aglaia (splendor), Euphrosyne (mirth), Thalia (good cheer)—often associated with Aphrodite.

*Hygiea:* A daughter of the god of healing and medicine, Asclepius, she represents health.

*Janus:* The double-headed deity, looking backward and forward at the same time.

*Justice:* Portrayed with scales and sword.

*Leda:* A beautiful maiden, usually shown with the swan, which is Zeus in disguise. Zeus came to Leda in the guise of a swan and from this union were born Castor and Pollux, the twins, and Helen, who became the cause of the Trojan War.

*Medusa:* The only one of the three Gorgons (others were Sthenno and Euryale) who was mortal. Poseidon fell in love with Medusa, who was once beautiful; she was turned into a monster with serpent hair by a wrathful Athena. Eventurally she was killed by Perseus.

*Meleager:* The main hero with Atalanta of the Calydonian boar hunt; depicted with spear and bow and quiver.

*Muses:* Most ancient works identify only three Muses, and their attributes are usually musical instruments, such as the flute, lyre, or barbiton, an ancient Greek musical instrument resembling a lyre. The three Muses are the daughters of Zeus and Mnemosyne (Memory). Later, nine Muses were associated with Apollo. Always shown draped and wearing long tunics, they are:

*Calliope:* Chief of the Muses and Muse of epic poetry. She is seen with a tablet and stylus; often with a roll of papers and a cloak around her waist.

*Clio:* Muse of history, represented with an open roll of paper or an open chest of books. Sometimes she carries two thongs and chastises one of the other Muses.

*Euterpe:* Muse of lyric poetry, represented with a flute or double flute or reciting lyric poetry.

*Melpomene:* Muse of tragedy, represented by a tragic mask; often carries a club or sword.

*Terpsichore:* Muse of dance and choral song, represented with the lyre and plectrum (used to pluck the lyre).

*Erato:* Muse of erotic poetry, also represented with the lyre.

*Polymnia:* Muse of the hymn, represented in a pensive or meditating pose with a roll and robe.

*Urania:* Muse of astronomy, represented with a globe and a staff to point out the stars.

*Thalia:* Muse of comedy, represented with a comic mask, carrying a shepherd's staff.

*Naiads:* Water nymphs that live in brooks, springs, and fountains.

*Nausikaa:* She befriended the shipwrecked Odysseus.

*Nereids:* The fifty daughters of Nereus, who live in the sea.

*Nessus:* A centaur killed by Heracles.

*Odysseus:* Called Ulysses by the Romans, the ideal hero of adventure, much beloved by Athena. King of Ithaca and a Greek hero in the Trojan War, his ten-year journey home from Troy is the story of Homer's *Odyssey.*

*Pan:* The woodland god of Arcadia, patron of shepherds and their flocks. He is playful, lascivious, and associated with the reed pipe.

*Paris:* A very handsome man, son of Priam, king of Troy, and queen Hecuba. His famous judgment of the most beautiful goddesses is mentioned in Homer's *Iliad.* He abducted Helen, which led to the Trojan War.

*Persephone:* The daughter of Demeter and Zeus and wife of Hades, god of the underworld. Persephone heralds the arrival of spring and is usually depicted with flowers.

*Providence:* A female figure, leaning upon a column with a cornucopia.

*Psyche:* The soul; identified by butterfly wings, she was the bride of Cupid.

*Satyrs:* Woodland spirits with horns, face of a man, body and legs of a goat; playmates of Pan and followers of Dionysus.

*Sileni:* Followers of Dionysus (Bacchus).

*Triton:* Son of Poseidon, a merman and trumpeter of the sea; his trumpet was a great shell.

---

Naturally some myths were more popular than others. The most popular ones from ancient times are used even today by contemporary carvers.

Athena was the goddess of wisdom who sprang fully grown from Zeus's head. Often accompanied by an owl, she became the patroness of Athens.

In the days of the Titans (the generation of gods that preceded the Olympian gods), the sea was ruled by Nereus, son of Mother Earth and Pontus, the boundless seas. Nereus was an old sea god with a long gray beard and a fishtail. He was the father of fifty sea nymphs, the Nereids; and a forerunner of Poseidon, the Olympian god of the seas and father of Triton, a merman mainly distinguished by his skill in sounding a conch shell (color plate 18).

Charon was the ferryman across the river Styx to the land of the dead.

The Fates, called Moirae by the Greeks, were Clotho, Lachesis, and Atropos (color plate 19). Clotho is a young woman, spinning man's thread of life. Lachesis is a woman in the prime of life, measuring the length of man's life. Atropos is an aged woman, ready with shears in hand to cut the thread of life.

The Gorgons were three hideous sisters, one of whom, Medusa, was killed by Perseus (color plate 20). Ovid, the Latin poet, recounted that Medusa was renowned for her beauty and roused jealousy in the hearts of many suitors. Most beautiful was her long hair, but when Neptune (Poseidon) seduced her in the Temple of Minerva (Athena), the goddess punished Medusa by turning her hair into snakes. Medusa was later killed by Perseus, who used her head to turn anyone who looked upon it to stone. The image of Medusa was considered a magic talisman and a potent charm; it was commonly featured on armor worn by Roman emperors.

The Medusa portrait provides an especially interesting dating challenge for jewelry appraisers, because it is found in both full face and three-quarter profile. The great jewelry artisan Fortunato Pio Castellani pointed out to C. W. King that the Medusa in profile, but not three-quarter face, is found in antique cameos. The image of the Gorgon as a living fury in full face with eyes staring wide dates from the Leonardo da Vinci painting of 1500. This must be disputed, however, because ancient Greek coinage showed Medusa in full face. Writer and glyptics expert John Boardman reports that the Medusa's looks were sweetened by the classical school of artists until at last she is shown as a lovely girl with snakes knotted neatly beneath her chin. Boardman notes that Hellenistic Medusas have full, fleshy faces, unkempt hair, and skimpy snakes.

It must have been very pleasant for the children living in Greece nearly three thousand years ago because so many of their questions were answered by fascinating stories. For example, if a child questioned the sunset, he was told that the sun is a splendid golden chariot that Apollo drives up the sky every morning and down again every evening. Even emotions were explained by myth. According to the Greek poet Hesiod, the love goddess Aphrodite was born of sea foam and floated to shore on a scallop shell, landing on the eastern coast of Cyprus in the Mediterranean sea. Pictures carved on stone illustrated and reinforced the myth. Aphrodite was called Venus by the Romans and was the object of cult worship on Cyprus. The amazing longevity of the myth is borne out by use of the subject much later by master artists Botticelli and Nicolas Poussin—both rendered celebrated paintings of the birth of Venus.

Most myths concern divine beings with human emotions such as love and jealousy (fig. 2-1). The Greek gods were very much like humans; some had likable qualities, some did not.

An interesting historical myth concerns the god Zeus and the queen of Crete, Europa. The story goes that Europa, playing in a meadow one day, was frightened by a beautiful but strange white bull that appeared next to her. The bull was so gentle, however, that she forgot her fear, hung garlands of flowers about his neck, and finally climbed upon his back for a

FIGURE 2-1.

Top: *The centaur Nessus, fleeing with his captive Deianira, Heracles's wife, intent on rape, is pictured in this agate cameo. A triton assists in the abduction.* Bottom: *An agate cameo showing Heracles in Olympus being taunted by the goddesses Athena and Aphrodite. (Photos courtesy of Achim Grimm)*

ride. The bull was actually Zeus in disguise (apparently something he did often), who suddenly galloped down to the shore, dashed into the water, and swam out to sea. He spoke to Europa to calm her fears and told her he was the god Zeus, come to earth to make her his bride because he loved her so much. He had stolen her away and would make her the queen of the island of Crete (fig. 2-2). When Zeus arrived in Crete with Europa clinging to his back, he put a royal crown of jewels on her head. Europa lived in Crete in riches and glory to the end of her days.

Explanation myths tell about natural phenomena, such as the formation of mountains, rivers, lakes, and oceans, weather and climate, and the seasonal changes. Persephone grew up on Olympus, the daughter of Zeus and Demeter, goddess of the harvest and fertility. Her mother loved her so dearly that she could not bear to let her out of her sight, for wherever Persephone danced, flowers sprang up. She was so beautiful that Hades fell in love with her and wanted her for his queen, and so he carried her off to the underworld. Persephone cried for her mother and longed for the sunshine. Above, on earth, Demeter searched for her lost daughter, and all nature grieved with her. Demeter withheld her favors of harvest and bounty until the earth was barren. Demeter called to Zeus that she would never make the earth green again if he did not command Hades to return Persephone. Since Zeus could not let the world perish, he sent

FIGURE 2-2.

*This chalcedony cameo, showing Europa and the bull, is a seventeenth- to eighteenth-century copy of a 100 B.C. onyx bowl fragment. (Content Family Collection of Ancient Cameos, Houlton, ME)*

word to Hades to let the girl go. But she was not to be parted forever from Hades and would return for several months a year to the underworld. And so it is that every year, when Persephone leaves the earth, nothing grows and there is winter, but as soon as her light footsteps are heard, the whole earth bursts into bloom and spring returns (fig. 2-3).

The often carved judgment of Paris concerns the cause of the Trojan War (color plate 21). As it is told in Greek stories, Paris, strikingly handsome, was selected by Zeus to judge which of the goddesses—Hera, Athena, or Aphrodite—was most beautiful. The prize was a golden apple. The three radiant goddesses standing before Paris vied for the award. "Give it to me," said Hera, "and all of Asia shall be your kingdom." "Choose me," said Athena, "and you shall be the wisest of men." "The most beautiful woman on earth will be yours if you give me the apple," promised Aphrodite.

Since Paris was so young, he admired beauty more than power or wisdom, and he gave the apple to Aphrodite. She took the apple without

FIGURE 2-3.

Left: *A late-twentieth-century interpretation of Persephone and Demeter enjoying springtime, carved in blue and white agate.* Right: *The goddess Artemis (Diana), identified by the crescent moon in her hair, with the huntress Atalanta, known by her quiver of arrows. This cameo presents an anomaly because of the presence of the chained dog. Though Artemis was protectress of all wild things, the dog was an attribute of the god Ares (Mars) and would not have been used in this manner by an ancient carver. Twentieth-century carvers, however, frequently incorporate ethnic folktale elements and their own imagery into carvings of classical myths. Both cameos were carved in Idar-Oberstein. (Photos courtesy of Achim Grimm)*

giving a thought to the fact that the most beautiful woman on earth was Helen, queen of Sparta, and already married. But when Paris sailed to Sparta to claim Helen as his own, she ran away with him without hesitation (with some fancy arrow work from Eros) and became Helen of Troy, the cause celebré of the Trojan War.

One of the most fascinating aspects of cameo study is identifying the subjects and designs and being able to understand and appreciate how the individual carvers have interpreted the myths and stories. Without some background or knowledge of the legends, some of the subtleties of the art form are lost, and the scene becomes just a meaningless jumble of figures.

Over thousands of years mythology has provided rich inspiration for painting and sculpture. Master carvers during the Renaissance and in the seventeenth and eighteenth centuries took elements from earlier paintings or bas-relief frescos as inspiration. Some elements — for instance the four winds — are often copied from a piece of classical art onto a cameo as a part of an entirely different scene, combining segments of several classical pageants to create a new picture (color plate 22). This kind of selectivity is prevalent in shell carvings today, in which the Italians not only resurrect the old classical myths, copy, and reinterpret them, but they also illustrate contemporary society, carving scenes that depict today's culture (color plate 23). In Idar-Oberstein, Germany, the modern interpretations of cameo art are most striking as abstract art (color plate 24), but carvers still draw inspiration from such masterpieces as the ceiling of the Sistine Chapel (color plate 25).

---

## SYMBOLS

In the *Handbook of Engraved Gems,* C. W. King explained the meanings of the use of numerous types of materials and their designs. He also quoted the ascribed virtues of various subjects from a publication dated 1502 and stated that because cameos from the Middle Ages were products of a lost art at that time the people believed the engravings were the work of nature, not man. The following list explains some common symbolism.

- A jasper stone engraved with a man with a shield in his left hand and an idol in his right, with vipers instead of legs and with the head of a lion, brings victory in battle and protects against poison.
- A jasper stone carved with a man with a bundle of herbs on his neck enables the wearer to diagnose diseases and stops the flow of blood.

- A jasper stone carved with a cross protects the wearer from drowning.
- A jasper stone carved with a wolf defends the wearer from snares and prevents the speaking of foolish words.
- A stag carved on any stone cures lunacy and madness.
- A lamb carved on any stone protects against palsy.
- A jasper stone carved with a virgin in a long robe with a laurel branch in her hand brings success in all undertakings.
- A jasper carved with a man holding a palm branch in his hand makes the wearer powerful and acceptable to princes.
- A carnelian showing a man seated and a woman standing before him with her hair hanging down to the thighs, casting her eyes upward, makes all who touch it obedient to the will of the wearer in all things.
- A green turquoise engraved with a portrait of Aquarius gives the wearer luck in all commercial transactions.
- A serpent with a man on his back and a raven over his tail, engraved on any stone, makes the wearer rich and crafty.
- Hercules holding a club and slaying a lion or other monster, engraved on any stone, will secure victory in battle for the wearer.
- Mars, on any stone, makes the wearer invincible.

Heliotrope from India, called bloodstone, was a favorite of early carvers, especially for modeling the head of Christ wearing a crown of thorns. The red splotches in the stone were made to represent drops of blood. According to an old legend, bloodstone was formed by the drops of blood of the crucified Jesus following the thrust of a Roman soldier's spear; the drops supposedly fell upon the green jasper on which the cross was standing. When the blood penetrated the stone, bloodstone was formed. From that time the stone seems to have been endowed by writers and storytellers with a divine power to stop hemorrhage from wounds; it was worn by Roman soldiers for this reason. The arrangement of red spots works equally well for other subjects, such as the Medusa the deadly Gorgon with wings and snaky hair, who was beheaded by Perseus with the help of Athena's shield and Hermes's sword (color plate 26). Medusa was a magic talisman worn as an effigy on Minerva's breastplate and was a common feature on the armor worn by Roman emperors.

In *The Book of Talismans,* author William Pavitt noted that in India today it is still customary to place a bloodstone upon wounds and injuries after dipping it in cold water. The curative qualities have some scientific basis, in that the iron oxide in the stone is an effective astringent.

## PORTRAITS

Early cameo carvers tended to produce portraits of living rulers or heroes, together with the symbols or attributes of a deity. Noble families liked this kind of presentation because most claimed divine descent or, if they did not, at least liked the comparison. Identification of the portrait was easy and definitive because of the inclusion of the symbols. Portraits from the fifteenth and sixteenth centuries, however, were generally clear, simple pictures uncluttered by superfluous effects. The heads alone might have classical features, but many are unidentifiable because they are individuals totally unknown today. An array of portraits of philosophers, scholars, heroes, kings, poets, inventors, and figures from the fields of architecture and literature are found in cameo portraiture of the fifteenth, sixteenth, and seventeenth centuries. How does one identify one face from among the hundreds of anonymous depictions? The answer is another question: how important is it to identify the face? For the average collector, it will be of little consequence; for the dealer and seller of cameos, it is more significant because if the portrait is of a famous person, the value may increase; for the antiquarian and jewelry historian, identification simply presents an irresistible challenge; and for the jewelry appraiser, greater professionalism and expertise is demonstrated if a portrait can be identified.

Cameo portrait identification can be approached in several ways. Because it has been a practice with carved portraits to use medals or coins as inspiration, look at early Greek and Roman coins in one of the myriad of coin books on the market, such as *Historic Gold Coins of the World,* to make comparisons. Also examine museum catalogs, catalogs of cameo dealers, museum fine-arts collections, antiquary cameo books (see the bibliography), public collections of cameos, and the illustrated biographies of philosophers, inventors, politicians, presidents, heroes, kings, queens, aristocrats, theater performers, and statespeople. Some books that contain numerous portraits are : *Dictionary of American Portraits,* by Jayward Cirker, with over four thousand pictures of important Americans from the 1600s to the start of the twentieth century; *The St. Memin Collection of Portraits,* by Gurney and Son, New York, with 760 medallion portraits of distinguished Americans from 1793 to 1814; *McGraw-Hill Modern Men of Science,* with 426 portraits of leading scientists; *Portrait Catalog,* G. K. Hall, a portrait catalog of 10,784 men of the New York Academy of Medicine and 151,792 portraits of medical doctors through the ages; and two books by Roy C. Strong, *The English Icon: Elizabethan and Jacobean Portraiture* and *Tudor and Jacobean Portraits. Portraits: 5,000 Years,* by John Walker, is a gallery of portraits of the notable from 3000 B.C. to the present. There are also portrait dictionaries of British painters and artists. One of the most important sources for portrait identification may be the books of portrait miniatures: *Miniatures: Dictionary and Guide,* by Daphne Foskett, and *Edwardian Portraits: Age of Opulence,* by Kenneth McConkey.

Sadly, after all the research, time, and effort invested in identifying a portrait on a carved cameo, it is entirely possible that the identity of the person will never be known because the portrait may simply be the likeness of an ordinary citizen of the era, unidentified and undistinguished from his peers, who had his countenance or that of a loved one carved in stone.

## Historical Events and Stories behind the Jewels

One of the most famous stone cameos in history is a sardonyx portrait of Queen Elizabeth I, set in a famous ring that she gave Robert Devereux, the earl of Essex, as a pledge of their friendship. The cameo was cut in three strata of color, mounted in a simple gold ring with enameled blue flowers on the back. When Essex was sentenced to death, he sent the ring to his cousin to deliver to Elizabeth, as a request for her to intervene in the sentence. Mistaking the cousin's sister, the countess of Nottingham, as the intended receiver, the messenger gave up the ring to a vengeful enemy of the earl. She did not deliver the talismanic cameo to Elizabeth as requested, and the earl was executed. Later, as the countess lay on her deathbed, she confessed the act to Elizabeth. The revelation so shocked and angered Elizabeth that, according to historians, the queen cried: "God may forgive you, but I cannot!" The ring today is owned by Lady Devereaux, a direct descendant of the earl.

Another love story concerning a cameo is told of a gold and enamel ring, circa 1720, now part of the collection of the British Museum in London. The ring is set with a ruby cameo portrait of Madame de Maintenon (1635 to 1719), a beautiful and resourceful young woman secretly married to Louis XIV in 1683. This was a period in France conspicuous for the waste of money and resources that had resulted from wars with the Netherlands. Madame de Maintenon was especially noted for her beauty and intellectual gifts, but she was unpopular with the king's subjects because she exerted strong religious influence over him and had considerable control over his foreign policy and domestic affairs. Raised in a convent, she became the favorite of Louis XIV when educating his children. After the death of Queen Marie-Thérèse, she quietly married the king, but refused to accept or use any of the crown jewels. The ruby cameo is said to have been commissioned by the king as her personal jewelry, and she wore it until Louis's death, after which she retired to the convent of Saint Cyr.

Not a cameo but an engraved gem all the same, the 71.70-carat diamond called the Akbar Shad was once part of an original-weight 116-carat diamond that contained two engraved inscriptions: "Shah Akbar, the Shah of the World, 1028 A.H."; and "To the Lord of Two Worlds, 1039 A.H., Shah Jehan." According to legend, this gem was one of the eyes of the Peacock Throne, built by Shah Jahan in Delhi and stolen by Nader Shah when he sacked Delhi in 1739. After 1739 the diamond disappeared, but it surfaced in Constantinople in 1866 under the name Shepherd Stone. It was bought by the London merchant George Blogg, who recut it into a 71.70-carat pear-shape stone, destroying the inscription. In 1867 the diamond was sold to the gaekwar of Baroda for $175,000 and has not been seen since.

The renowned stage actress Sarah Bernhardt had a unique relationship with the great artist and designer René Lalique that can be traced to 1890, when she posed for a life-size bas-relief later reduced in scale and struck as a presentation metal. Lalique was commissioned to plan and design the stage jewelry for some of her most famous performances: *Iseyl, Gismonda,* and *Théodora.* Although accustomed to gifts of gold and precious stones from her many affluent admirers, the actress favored above all other items of jewelry a belt of hardstone cameos that was given to her by some gold miners during a visit to California.

An extraordinary cameo emerald necklace made in the early nineteenth century was auctioned by Sotheby's in Geneva in 1990. The presale estimate of its value was $400,000. The piece was commissioned as a gift for Queen Maria Carolina, wife of Ferdinand I of the Two Sicilies and sister of Queen Marie-Antoinette of France. The necklace contains five large oval emerald cameos carved with classical portraits, each in a setting surrounded by diamonds. Fascinating provenance adds interest to the necklace: it was given in payment for public works in the Kingdom of the Two Sicilies to Josef-Michel Montgolfier, the celebrated balloonist. One hundred years later it was returned to the Bourbon family through the marriage of Prince Gabriel to Cecile Lubomirska, a descendant of Montgolfier.

That cameos have been highly prized for ornamenting crowns and royal diadems is borne out by the many existing examples still intact in museums worldwide. In earlier days crowns of royalty were subject to numerous reworkings of their style and decoration as rulers changed. The European monarchs were often unable to maintain the crowns in full splendor, and frequently diamonds and other precious gemstones were rented from the court jeweler, then returned after the coronation. What was passed down, however, and remained on the gold frame of the crown, were many historic cameos from earlier eras. A great many ancient hardstone cameos were mounted on the crown of Richard earl of Cornwall, when he was crowned king of the Romans in 1257. Napoléon specifically requested antique cameos for his crown, which he later called the crown of Charlemagne. One of the most exquisite examples of cameo encrustation in a diadem belonged to his empress Joséphine. It is a gold diadem mounted with large framed shell cameos, richly augmented with precious gems. The cameos all show mythological scenes: the large center oval cameo depicts Apollo in his sun chariot; the other cameos portray cupids in a variety of occupations, in the style of the Pompeii frescoes. In Russia too cameos were used in the crown jewels. Ivan IV was crowned as the first czar of all the Russias in 1547; Peter the Great became czar in 1689. Both used carved and engraved gems in their royal regalia.

Engraved gems were passionately collected by Pope Paul II (1464-71) who was prepared to pay almost any sum for them. One story about his obsession for collecting cameos concerns the city of Toulouse. When he learned that the city owned an outstanding ancient cameo of great beauty, he not only offered the city a huge amount of money for it but extra privileges for their basilica of Saint Saturin and a new bridge into the town! The city did not accept his offer.

Of the truly outstanding collections in extant is that originally of William, third duke of Devonshire, acquired during the first half of the eighteenth century in London. This incredible array of cameos was delivered into the hands of jeweler C. F. Hancock in 1855 to make into a complete parure of seven pieces of jewelry. The set, known as the Devonshire Parure, consisted of coronet, diadem, bandeau, comb, necklace, stomacher, and bracelet, all of gold and set with diamonds plus eighty-eight ancient cameos and intaglios. The entire suite of jewelry was worn by the countess of Granville, wife of the British minister to the court of Russia, at the coronation of Czar Alexander II in 1856.

The central cameo in the diadem was an exquisite antique of Victory in her chariot cut on a gray, brown, and white onyx in low relief. On the back of the cameo, a fifteenth-century artist had carved a cameo of a river god. Other cameos in the diadem included several portraits of Queen Elizabeth and one of the earl of Leicester, in all about twenty pieces. The comb, encrusted with cameos of various sizes, had one with a head of Leander, one with a centaur with a bacchante on his back, and one with a faun. The bandeau's most significant cameo was a large emerald carved with a full face of Medusa. The necklace had sixteen cameos mounted as pendants with portraits of Queen Elizabeth in high relief, Edward VI, and Tiberius. The stomacher was mounted with sixteen antique cameos of the Renaissance, including one of Europa and one of Minerva. The bracelet was set with numerous small cameos. This collection, one of the most important ever brought together, included gems of worldwide fame. Sadly, it was broken up in 1899 and is now in private and museum collections.

Finally, of the cameo encrusted boxes and vessels in museums in the United States, an especially interesting Spanish leather box set with cameo portraits and battle scenes resides in the Walters Art Gallery. Of historical importance, it once held the pennant of King Frances I, taken after his capture at the battle of Pavia, February 25, 1525. On that date the king of France was defeated by the Spanish knight Don Juan López Quixada. According to the rules of chivalry, the captive surrendered all his possessions, including his arms and armor, to his captor. López Quixada, in service to Emperor Charles V, offered the symbols of victory to his lord and was allowed to keep the captured battle pennant. The knight later had the box made as a container for it, with a proud inscription proclaiming that it was he who had captured the standard of the king of France.

### Instant Expert

***The Twenty Most Common Subjects of Cameos***

The following list, in no particular order, represents the twenty most common subjects found on the revival cameos of the nineteenth and twentieth centuries.

Zeus

Athena (Minerva), usually helmeted

Baccante maiden (cult of Dionysus or Bacchus) with vine leaves and/or grapes in her hair

The three Muses

Nike, sometimes known as Victory, or Fortuna driving a two-wheeled chariot

Medusa

Portraits of men or women, may be gods or goddesses, sometimes carved two together in conjugated form

Eros (Cupid) alone, in multiples, or with animals

Leda and the swan

Ganymede feeding the eagle; sometimes Hebe feeding the eagle

Birth of Venus, or Venus rising from the sea

Helmeted warrior with a dragon carved atop his helmet, meant to be Saint George; sometimes interpreted as Mars

Cupid riding a lion, interpreted as love conquers all

Artemis (Diana) with a bow and quiver

Venus and Cupid

Judgment of Paris

Cupid in chains, interpreted as love bound

Rebecca at the well (biblical reference)

Angel and child (biblical reference)

Profile of an anonymous woman (mass-produced in stone and shell)

*C h a p t e r*

# 3

# CAMEO PRODUCTION

## TOOLS AND TECHNIQUES

The tools used by ancient gem craftsmen before the Iron Age were made of stone, bronze, and copper. Bow-driven points of soft metals, such as copper, and hand-rotated drills were known by Mesopotamian carvers as early as 4,000 B.C. Knowledge of their use reached European carvers between 1800 and 1600 B.C. The earliest drill was a hand-held vertical shaft tool with a piece of flint attached and used as a bit. A bow was drawn back and forth across it, causing the rock bit to rotate. C.W. King suggested that the rock bit was changed to a metal point and cutting wheels were used in about 2000 B.C., when craftsmen began carving harder materials such as chalcedony. The methods for drilling, cutting, and polishing remained virtually unchanged until the introduction of electric power.

When early carvers began work, they fixed the stone in a bed of some type of cement, and the tools were held by the engraver (in contrast to modern methods, where the drills and wheels are fixed and the gemstones are loose in the hands). Pliny records that hardstones such as quartz were carved or engraved using an abrasive paste of emery dust in oil or water. The development of sapphire or diamond splinters mounted in iron or bronze handles for fine-line finishing provided the carver with an opportunity to vary engraving techniques. Polishing was accomplished with metallic oxides or soft woods.

In carving or engraving, the major portion of the work was done by the friction of the diamond or corundum powder mixed in oil and continually fed into the rotating point. The entire carving process was dependent upon the abrasion, with the abrasive powder being the most important ingredient. This fine powder, mixed with oil, was kept smeared on the drill; thus the tiny particles clung to the metal under pressure, and in that way a steady cutting surface was provided.

The intriguing question of how the ancient carver managed such technical and detailed cameo carvings without the aid of some type of magnifying glass is often answered unsatisfyingly—the craftsmen were myopic or had unusually close vision.

"It would be difficult to believe," noted Gisela Richter in her *Catalogue of Engraved Gems of the Classical Style,* "that the ancients could execute the minute work they did without lenses." But as an afterthought, she added, "Nowadays, when strong lenses are available, gem engravers do not always use them." However, she believed the ancient craftsmen probably did use them and supported the view with numerous first century A.D. writings on the use of balls of glass. One highly important observation came from the Roman rhetorician Seneca (55 B.C. to A.D. 39), who recognized the ability of glass to magnify and wrote: "All objects appear much larger if seen through water; letters, however minute and indistinct, appear larger and may clearly be seen through a glass full of water."

Debate as to whether or not glass or optical clear quartz lenses were used as magnification aids has been waged among scholars for centuries, and there is still no clear consensus. The Romans did make lenses from glass; one was discovered in 1854 at the House of the Engraver in Pompeii. Two lenses that have been dated A.D. 174 were found at the home of a Roman artist at Tanis in Egypt.

Lenses were used in several ways, as a means of kindling fire, for example; but numerous lenses have been excavated that seem to have had other purposes, probably magnification.

It is generally agreed among researchers that the fine details of seal stones would have required careful scrutiny to distinguish them from forgeries, and this type of close observation would have required some magnification. Dr. George Sines of the University of California has been studying the issue and believes there is convincing evidence that lenses were used for magnifying purposes, although it may not have been a common practice. In a 1987 paper that Sines co-authored with Yannis A. Sakellarakis about a lens excavated in Crete in 1983, Sines noted the glass had "fine optical quality." He recounted how his interest in the subject led to an investigation that showed the use of lenses was widespread throughout the Middle East and the Mediterranean for thousands of years. During the course of the research, Sines found a cache of twenty-three ancient lenses displayed in the Archaeological Museum at Herakleion in Greece with countless more examples in museum storage. An interesting aspect of many of the lenses is the rock crystal optical quality and the fact that most had a useful magnification factor of at least seven times. Pointing out that the lenses were plano-convex, Sines wrote that they showed very shallow, slight circumferential tooling marks that were consistent with using a cutting stone to shape the periphery, as well as a template. Another argument for the use of lenses, according to Sines, is the standard spacing found in the cable borders on the carved gems. This fact alone, he believed, is strong evidence of use of a lens that magnified four to ten times.

The question has puzzled many over the centuries, and Lorenz Natter, an eighteenth-century carver, became convinced after extensive research on the subject that the ancient Greeks used magnification: "The art of engraving in gems is too difficult for a young man to be able to produce a perfect piece; and when he arrives at a proper age to excel in it, his sight begins to fail. It is therefore highly probable that the ancients made use of glasses, or microscopes."

---

## MODERN METHODS

Today the power lathe is the most important tool employed by carvers. They use a fixed motor with a shaft holding a small chuck. The bits and cutters are usually impregnated with diamond dust and are mounted into the chuck. The cutters, or burrs, have a variety of shapes in various designs: wheels, ovals, cones, cylinders, and so forth (color plate 27). When a cameo is hand-carved, the artist holds the material in his hand (color plate 28), or cements it to a dopstick and moves the stone against a rapidly revolving cutter (color plate 29). Normally the carver first prepares a cameo blank by sketching a design, portrait, or scene on the surface of the stone with a diamond stylus. Then the carver carefully cuts the superfluous material away from the design, which is constantly refined. The details are made sharper by using increasingly fine tools and burrs (color plate 30). Cameo carving on shell is handled with some differences: the execution of the design today is still largely accomplished using only hand gravers supplemented with a small dental drill (color plate 31).

A modern carver has a large assortment of tiny saws and carving points, some fine, some broad, like knitting needles (color plate 32). A cameo cutter will begin work on his previously prepared blank of stone by first drawing the design, freehand, from a sketch, photograph, or picture onto the stone. All the superfluous layers, down to the background layer, have been removed with a saw. The design is drawn using the largest drill points, gradually working down to finer points until the lathe is no longer used and work is continued using a diamond point. All finishing touches are done with a diamond point. Polishing a cameo is an important but tedious process, because to show detail the stone must be well polished. Because polishing can destroy detail, finishing and polishing are done alternately until the carver is satisfied with the design and then single lines are polished, one at a time. Most carvers begin to polish their work even before they are finished with the design, cutting the final lines into the polished surface, and then polishing these.

The cameos are usually held in the hand or attached to a dopstick, especially when being polished. The materials used for polishing are wood or soft cloths fed with oil and mixed with polishing powders such as tripoli, alumina, oxide or iron, or rouge. With some hardstones, the carver may choose diamond dust and oil to polish his creation. One of the greatest German cameo carvers, August Wild, was said to have used his tongue and olive oil to achieve an outstanding polish on his gems. His unorthodox polishing method was a highly guarded secret that he revealed only shortly before his death.

While some hand carving on hardstones is still carried on in Idar-Oberstein, Germany, in the traditional way, most work is accomplished with the help of motors, arbors, rheostats, and a wide assortment of burrs, cutting and sanding discs, and polishing tools. And now there is a challenger to the skills and traditions of the cameo craftsmen: the ultrasonic carving machine (fig. 3-1).

---

## Ultrasonic Carving

The ultrasonic carving machine has the ability to mass-produce hundreds of cameo clones from one casting in a matter of minutes, in contrast to the hand-carving of a medium-sized cameo, which can take many hours, sometimes days. The new technology has revolutionized cameo production for the mass market; this is an important milestone in cameo evolution, the twentieth-century legacy to the art. Ultrasonic cameo carving is being conducted today in Germany, Japan, and in the Pacific-rim countries.

Most German cameo carvers look pained when questioned about the ultrasonic method of cameo production: they would prefer that the entire secret stay safe and secure in the workshops of their Idar-Oberstein businesses. Though the Japanese are less secretive about the technique, they share information only to a point and then will tell no more. The process, however, has been explained by an American manufacturer and distributor of the ultrasonic carving machine, who provided detailed and

**FIGURE 3-1.**

*Ultrasonic stone-cutting and cameo-carving machine. (Photo courtesy of Sonic-Mill, Division of Rio Grande, Albuquerque, Inc.)*

valuable information; the following account of the machinery and how it works regarding cameos is described here.

The American manufacturer, Sonic-Mill in Albuquerque, New Mexico, distributes the unit worldwide (fig. 3-2). The machine is suited for high-volume application, operating automatically once the parts are loaded. There is a programmable controller, a microprocessor, for ease of production. A power supply converts conventional line voltages to 20-kilohertz electrical energy. This high-frequency electrical energy is relayed to a piezoelectric converter, where the electricity is changed into mechanical energy. The motion from the converter is amplified and transmitted to the horn and cutting tool, causing the horn and attached cutting tool to vibrate perpendicularly to the tool face thousands of times per second without any side-to-side motion. A recirculating pump forces an abrasive, suspended in a liquid medium, between the vibrating tool face and the work piece. These abrasive particles strike the work piece at one hundred fifty thousand times their own weight, chipping off microscopic flakes from it and

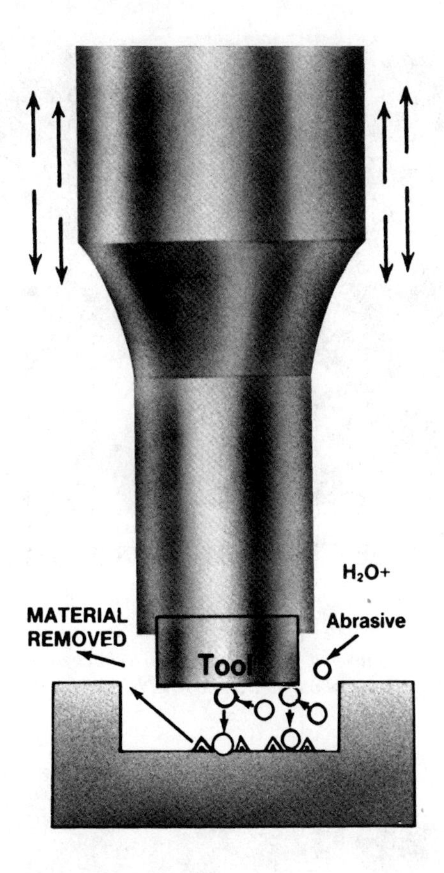


FIGURE 3-2.

*A schematic showing how the Sonic-Mill cameo-carving unit operates. (Photo courtesy of Sonic-Mill, Division of Rio Grande, Albuquerque, Inc.)*

grinding a counterpart (replica) of the tool face. Because there is no direct tool-to-work contact with a cool slurry, this is basically a cold cutting process.

To begin work with an ultrasonic machine, a casting is first made in metal, usually brass, from a stone prototype (a hand-carved cameo or an original piece). Or the prototype may be made with a regular wax injection casting in a rubber mold. After the metal casting is made, it is mounted onto a chuck and put onto the horn of the machine (fig. 3-3). A prepared stone blank is mounted onto a metal plate with wax; because of the electromagnetic properties of the chuck, the blank adheres to the plate. The machine is turned on, and moments later the cameo is finished. The brass castings used on the horn have a short life, producing only about

FIGURE 3-3.

*After the metal casting is completed, it is mounted onto a chuck and put onto the horn of the machine. One casting yields about ten cameos. (Photo courtesy of Pointers Gemcraft Manufactory)*

FIGURE 3-4.

*The Pointers Gemcraft Manufactory ultrasonic cameo machine. This company makes cameos in Hong Kong and mainland China. (Photo courtesy of Pointers Gemcraft Manufactory)*

ten pieces per casting, but since the castings are very inexpensive—about twenty-five cents to one dollar each—production costs are considerably lower than for hand-carved cameos.

A. Ruppenthal, a company of cameo carvers in Idar-Oberstein, reports that as much as 80 percent of cameo production is now done ultrasonically. They note that agate is the best material for use in the process, which is broken down into four basic steps: cutting the blank stone to size; dyeing the agate; ultrasonically processing it; and finishing by hand.

Pointers Gemcraft Manufactory of Hong Kong and mainland China supplies the market with an ultrasonic cameo carving unit that looks similar to the Sonic-Mills unit and works on the same principle (fig. 3-4). The Pointers spokespeople confirm by photograph that numerous ultrasonically carved pieces are hand-finished to make them more salable (color plate 33).

## Distinguishing Ultrasonic from Hand-Carved Cameos

The majority of collectors today do not even know that ultrasonically carved cameos are being produced. And, among those who do know, how many actually care that the addition to their collection was crafted by a machine instead of a master carver? Some do care. Collectors, if given a choice, want originality along with creativity and exclusivity. Since ultrasonically made cameos are clones of an original model, it is difficult to tell the copies from the original, but it can be done. Most manufacturers, dealers, wholesalers, and retailers of cameos who know how to differentiate the original from the replica will point out that the clones have machinelike precision and complete lack of undercutting. *Undercutting* means that the bas-relief is cut away from the background, slightly or deeply, sometimes producing an almost three-dimensional design (color plate 34). Undercutting cannot be done by a machine; however, it *can* be found on a machine-carved cameo if the manufacturer has taken the time and trouble to have the cameo hand-finished, adding undercuts to make it look more like a hand-carved piece. Because mass-produced cameos are very inexpensive, however, and hand-carving is a labor-intensive process, manufacturers usually will not go to the trouble or expense of hand-finishing ultrasonic cameos. Of course, when one looks at enough cameos, a complete uniformity of color, subject, and design becomes apparent. If you see the same female profile carved identically in hundreds of blue-background or brown-background agates, you must assume it is mass-produced, ultrasonically.

There is an absolute and unmistakable production giveaway on the ultrasonically carved cameo that you can learn to spot. After many hours of research on this subject in Idar-Oberstein and endless consultations with carvers and master craftsmen, the key to distinguishing ultrasonically made cameos was discovered. Color plates 35 and 36 are magnified close-ups of an ultrasonically carved cameo of a woman's profile in dyed brown and white agate. The cameo is approximately 18 by 25 millimeters (⅔ by 1 inch). Observe the surface portion of the white layer on the face, and consider how the ultrasonic machine works: a vibrating process, in which grit is propelled from the stone in a chipping-away procedure (color plate 37). That procedure should leave some kind of machine track, and it does. Close and long observation of the white layer shows it has perfect symmetry and looks shiny and compact. It has a very special kind of glisten, which can best be described as the look of fresh-fallen snow. This descriptive phrase was coined by Gerhard Becker, a German master cutter.

The very slightly pitted look is so unmistakable that it is surprising

FIGURE 3-5.

*Pictured are two hardstone cameos. The one on the left, black and white onyx with the profile of a Grecian woman, is a hand-carved nineteenth-century German piece by Hermann Grimm. The cameo on the right is an ultrasonically carved blue and white agate produced in Idar-Oberstein. Visual differences are immediately apparent, even in photographs. (Hand-carved cameo courtesy of Achim Grimm)*

that it has escaped detection for such a long time. Since the ultrasonic process was developed in Germany over two decades ago, consider how many hundreds of cameos with this provenance must be in the hands of collectors!

The fresh-fallen snow (FFS) syndrome is easily seen with the aid of a ten-power jeweler's magnifying loupe (figs. 3-5 and 3-6). Once recognized, it is never forgotten. A master wood craftsman discussing the ultrasonic process compared its look to that produced by sandpaper, instead of a hand scraper, in finished wood products. Another noticeable difference is in the feel of the ultrasonic. If one has a sensitive touch, an ultrasonically carved cameo will feel slightly resistant when rubbed with a thumb, while the hand-carved one will feel silky smooth.

FIGURE 3-6.

*A close-up view of the ultrasonically carved cameo shown in figure 3-5.*

Even ultrasonically carved but handfinished cameos show the FFS syndrome on some portion of the piece and can be distinguished from those cameos that are totally hand-carved. To gain expertise in this identification technique, many hardstone cameos should be inspected. Buy several from a jewelry retailer who knows which part of the stock has been ultrasonically carved and he or she will point the stones out to you. When you have the cameos in hand, study them tirelessly until the tracks of the ultrasonic technology are as clear to you as your own mirror image. These tracks are the fingerprints of modern technology.

Another clue in identifying replicas is the sales price of the cameo. Those carved in agate and other hardstones by hand are expensive because labor and materials are no longer cheap, and one must also pay for creative genius. It is logical that if a hardstone cameo of a moderate size is marked with a surprisingly low price, it may be an ultrasonic clone.

A large quantity of second- and third-quality German and Japanese cameos are being marketed in the United States. Once the lesser-quality cameos are set in gold mountings, flaws are well concealed, and the pieces are attractive enough to the average buyer, who knows nothing about distinguishing quality or machine carving. The second- and third-quality ultrasonically carved cameos are even less costly than their first-quality counterparts. To identify them, look for the following blunders: uneven and outrageously crooked borders; noticeably uneven thickness of the layered materials (you can see this from a side view of the cameo); out-of-oval or out-of-round shapes that were intended to be perfectly formed; designs that spill into the border unintentionally; poor or incomplete modeling of the figure or portrait; a white layer so thin it allows the color background to show through it; no clean and clear distinctions of white-layer designs from the background material. These cameos are not art; they are machine mistakes.

---

## MATERIALS

### HARDSTONES

There is little difference between the materials used during the Renaissance and those preferred by today's artisans. The major distinctions are the abundance of new materials and the conception, then and now, of what can be accomplished with the material. Early craftsmen were not particular about the quality of the stone they carved and paid little attention to the arrangement of the strata. Some early cameos have the design carved perpendicularly into the layers, instead of the more artful horizontal cutting into the colored strata. Single-colored stones were used but not as favored for cameos after the introduction of the banded Indian sardonyx. The reason was simple: cameos carved in one-color stone lost the prized bas-relief effect, or at least it was not as pronounced. Some single-color stones were cut into cameos because the material was highly valued, for instance lapis lazuli and malachite; sometimes the precious stones diamond, ruby, emerald, and sapphire were used.

Lapis lazuli, which came from Persia and was of a deep blue color, was extremely valuable to the ancient carver, as well as to the artist, who used blue pigment derived from the stone that he called cyanus, or ultramarine. This magnificent blue was quite unrivaled in color and durability. It is found on many ancient frescoes of the medieval period, when it was commonly sold for its weight in gold. Most of today's finest lapis lazuli comes from Afghanistan, and it is still a favorite of gemstone carvers.

Opal is a relative newcomer to the carver's stock of materials. William Schmidt, an English carver, documented that he was the first to carve the rainbow-colored material into cameos in 1874. He wrote in a letter to A. Booth that he invented a process of cutting opal cameos in such a way as to utilize the matrix of the rough opal for the ground color and emphatically insisted that no antique or fifteenth-century opal cameos existed.

The cameos Schmidt carved were exhibited in the 1878 Paris Exhibition by John Brogden and received a gold medal. They were acclaimed by the Paris *Daily Telegraph* in a June 12, 1878, news story about the exhibition:

*Mr (John) Brogden shows a neck ornament which in its nature is unique. The gems in this beautifully designed piece of decoration are camei, each being cut from an opal or the matrix of an opal. Now, to appreciate the delicacy of such carving, it must be remembered that the iridescence of this curious stone is due to the minute fractures by which its entire substance is traversed, just as though it had been shivered by some natural shock. As may well be imagined, the splintered formation of the opal renders it an exceedingly impracticable stone to cut; and though here and there a carved or engraved opal may be met with, this kind of workmanship is so rare that a cameo necklace must be accounted a wonder of wonders in the jeweller's art.*

Diamonds, rubies, sapphires and emeralds have never been used in any great quantity for cameo carving, although there have been notable exceptions. The materials are not only scarce and expensive, but also much more difficult to work because of their hardness properties. One of the few recorded examples of a diamond cameo was the one owned by Pope Julius II (1443 to 1513). Engraved with the figure of a friar, the cameo was executed by carver Ambrosius Caradossa. No record of the cameo's size exists, and its whereabouts today are unknown.

During the art deco period of jewelry design, green and blue color combinations became important as a new "Oriental" way of combining colors. Sapphires and emeralds were engraved and carved with Mughal flowers on a necklace commissioned by the Aga Khan in 1927. There is no record of the present whereabouts of the Cartier-made necklace.

Quartz has long held an attraction for carvers, especially the purple quartz called amethyst. Amethyst was a favorite carving material of the Greeks and Romans, and the stone today enjoys popularity among carvers in Idar-Oberstein because of its soft colors and translucency, presenting a challenge of carving a memorable cameo.

The materials most consistently sought by cameo carvers are the quartzes known as *chalcedony*, which are distinguished by their minute crystalline structure, visible only under the microscope. The term *chalcedony* refers specifically to white, gray, blue-gray, and black varieties. Other varieties, which differ in color because of the presence of impurities, are carnelian, sard, chrysoprase, bloodstone, moss agate, agate, onyx, jasper, and sardonyx. Carnelian is a low-intensity red to orange color that is semitransparent to translucent. Sard is brownish-red to red-brown translucent chalcedony, darker in color than carnelian. Chrysoprase is a yellowish-green translucent stone. Bloodstone is semitranslucent dark green with reddish spots. Agate is a translucent chalcedony with curved or irregular bands of colors. Onyx is the name given to the agatelike chalcedony in which the bands are straight and parallel. Jasper is an opaque chalcedony that occurs in single colors of red, yellow, brown, or green, or in a combination of colors.

Sardonyx is onyx with alternate bands of sard or carnelian with white or black layers. Sardonyx was mentioned as a gem of great value by the Romans, who believed its use as a material for signets was critical. It is interesting to note that Socrates classified the worthless, idle men in his society as "sophists, soothsayers, doctors, weather prophets and lazy long-haired onyx-ring wearers." If anything, sardonyx became even more fashionable five hundred years later, when a Roman lawyer, to be successful, had to live extravagantly and to appear prosperous often rented a costly sardonyx ring.

As noted by Jack Ogden in *Jewellery of the Ancient World,* the study of materials is often clue to discovering frauds. Some so-called Roman gems are materials traceable to South American mines, totally unknown to the ancients, Ogden said, while others have been proven to be made from modern synthetic stones or gems unused by the ancient carver.

---

## Color

Ancient craftsmen, as well as those of some later times, believed that engraving certain colored stones with specific designs produced a magic power capable of bestowing health and good fortune or of protecting the wearer against accidents. To the ancients, color was no mere decoration. Related to unseen and mysterious forces, color was tied to the sun, sky, earth, stars, and rainbow; it had a mystery and magic all its own. When scenes, portraits, and myths were carved in certain stone materials, they had their magic reinforced; to have the material in a specific color amplified the enchantment.

Today the most important of the carvers' materials is unquestionably the chalcedony group. In Idar-Obserstein, where the demand for this material is high, the craftspeople know how to stain the stones artificially with oxides and metallic salts to obtain desirable colors. All natural-colored varieties of chalcedony get their color from impurities and trace minerals. Iron oxides cause brown in agate and sardonyx and red in carnelian. Nickel is the cause of the green in chrysoprase. Natural black onyx is rare; most black material is produced by soaking in a sugar solution and then heating in sulphuric acid to carbonize the sugar. The Romans used a similar process.

Many publications give directions for treating onyx to produce specific colors. Cyril Davenport's *Cameos* advised: "Reds by means of pernitrate of iron. Black by oil, honey, or sugar. Blue by iron with ferrocyanide of potassium. Green by nitrate of nickel." Often high and sustained heat is enough to improve the color of onyx or agate, a process mentioned by Pliny.

Dr. Kurt Nassau has made a dedicated study of the process of coloring chalcedonies and other stones and outlined the procedures in his book *Gemstone Enhancement.* He noted that many porous gemstones can benefit by dyeing or staining, especially chalcedony, agate, carnelian, and onyx. One of the simplest dyeing processes, Nassau said, is the use of muriatic acid to dissolve the small amounts of iron oxide found in most of the stones. The second step is heating the stone, causing the iron to distribute on the internal surface of the stone, yielding the sought-after yellow-brown or red color. Nassau stated that if some iron salt, or even an iron nail, is added to the acid, the color is intensified. A large number of pamphlets and books on coloring stones have surfaced in the last decade. Once a secret process, it began to be revealed to the public by such noted mineralogists as John Sinkankas, in his book *Gemstone and Mineral Data Book,* with detailed instructions about dyeing gemstones taken from his extensive research and personal observations. A few of the publications about coloring gems have come from Germany, where the activity is as vigorous as ever. However, since many of the authors risk the displeasure of their colleagues by disclosing the secret methods of stone treatments, vital information and preparation steps have sometimes been deleted from the "recipes."

A great many cameos have been carved from glass, and the colors of the ancient glasses were especially bold: ruby red, sapphire blue, emerald green, orange-yellow. The colors in the glass were produced by the use of various metallic oxides and salts. Blue, green, and ruby red were produced by mixing different oxides and salts of copper. An amethyst-purple color was achieved with the use of manganese, blue from cobalt, yellow from carbon. The opaque white that is seen in the white stratum of imitation onyx was produced with oxide of tin. Pliny mentioned the use of oxides for color in glass and said that the chief distinguishing features between pastes, or glass, and natural gems were the hardness, weight, and coldness of the genuine gemstones.

---

## SHELL

While they may have been carved earlier, cameos are believed to have been first cut in shell around the fifteenth to sixteenth centuries. If they were carved earlier, they may not have survived, because of the nature of the material. Earliest specimens of shell cameos appear to have been carved from some type of mussel or cowrie shell, giving the relief portions of the design a gray color. Shell cameos were popular in France in the nineteenth century; Empress Joséphine delighted in them.

Queen Victoria of England (1819 to 1901) was fond of the shell cameo and made its wear fashionable in the nineteenth century. In the mid-Victorian period, when clever craftsmanship was appreciated, cameo *habilles* were made: the portrait head was shown wearing a necklace and earrings, which were set with tiny diamonds. This type of cameo is still being made today.

The shells most commonly used by the carvers are *Cassis rufa* and *Cassis madagascariensis.*

The *Cassis rufa* (bull mouth), from the East African coast, averages from 5 to 8 inches in size. The layers of the shell vary in color from a very light orange to a deep brown-orange. The *Cassis rufa* is known as the carnelian shell.

The largest mollusk used for cameos is the *Cassis madagascariensis* (emperor helmet), often 12 or more inches long, with a thick outer wall and a dark brown interior. Known as the sardonyx shell, a cameo carved from this shell looks like sculpted marble (color plate 38). The shell is native to the eastern coast of lower Florida; it found its way to Italy via the Bahamas. Some one hundred years ago, it was mistakenly thought to be native to Madagascar, thus its name. Because of the size of the shell, elaborate scenes are often carved upon it that are complete in themselves; the cameo is not cut away from the shell (color plate 39).

Other shells used for cameo carving are the *Cassis tuberosa* (helmet), found in the West Indies, with colors of white and brown, and the pink and white *strombus gigas* (pink queen's conch) from the Bahamas, a very popular shell but not very satisfactory for a carved cameo because the colors are subject to fading.

Their soft, muted colors distinguish shell cameos from hardstone ones. Some cameo purists argue that only hardstone carvings can rightfully be called cameos and prefer to call shell cameos bas-relief works. Be that as it may, 95 percent of all the cameos sold in the world today are the work of Italian craftsmen working in shell. According to the chairman of the Italian Association of Manufacturers of Corals and Cameos, the major purchasers of cameos are the United States and Japan. The figures released for 1989 by the association reveal that sales of cameos totaled about $270.4 million.

Most cameo shells have three layers of color (fig. 3-7). First is the outside layer, sometimes warty in appearance, which is usually cleared away, unless it is to be used in some specific design. The middle layer is generally white and is the layer upon which the design is carved. The third layer provides the ground color, as the contrast for the middle layer.

The cameo has been produced in Italy since the Roman artists engraved gemstones for their emperors and rich patrons. Rome became the center of the glyptic arts and remained so through the medieval period, when popes and cardinals served as patrons of the arts. After the fall of the Roman Empire, some glyptic artists settled in Naples and continued their trade of engraving seals for merchants and others, using agates and other hardstones. These materials were costly, and when they became too difficult to get, carvers were forced to search for substitutes. The one that could replace agate and at the same time retain its intrinsic beauty was found in the exotic multicolored shells of the east African coast. The shells were

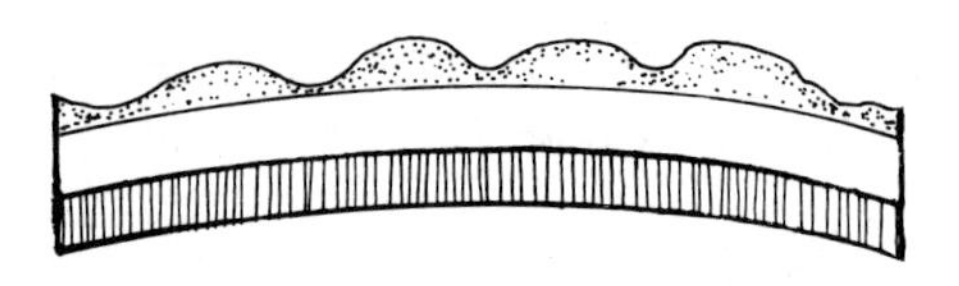

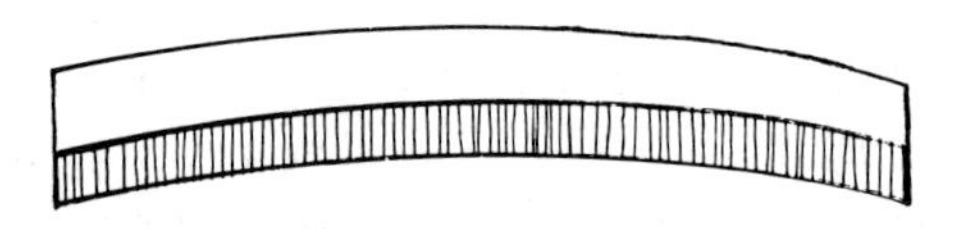

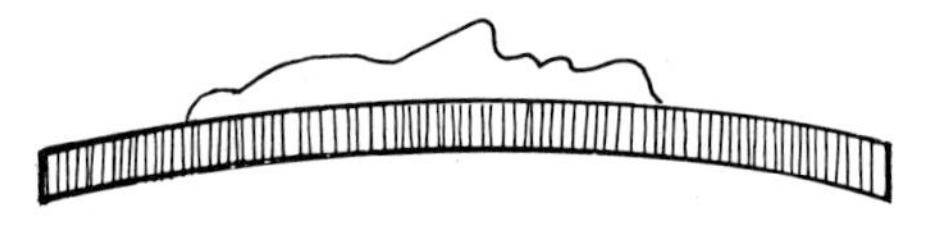

**FIGURE 3-7.**

*Most cameo shells have three layers of color. (Illustration by Elizabeth Hutchinson)*

brought to Naples by fisherman, who allegedly carved them to pass the time on their long voyages home. When carvers saw the shells, the acceptance was immediate. They praised the material for its easy carving qualities as well as super abundance, which helped lower the cost of production.

The steps involved in producing the shell cameo, from importing the raw material to producing a finished piece of art ready to be set in a mounting, are similar for all manufacturers.

The selection of the shell to be cut is the first consideration (fig. 3-8). The *Cassis rufa,* or carnelian shell, with its yellowish, warty outer crust, is a popular choice.

Step two divides the shell into two predetermined parts (figs. 3-9 and 3-10). The separation is made with a 10-inch diamond-edged disc blade, operating in water. One part will be used to make souvenir items of lesser value (fig. 3-11). The other part, called the cup, is marked for cameo shapes (fig. 3-12). The cup contains the thickest part of the shell and a richly colored background, which gives color contrast.

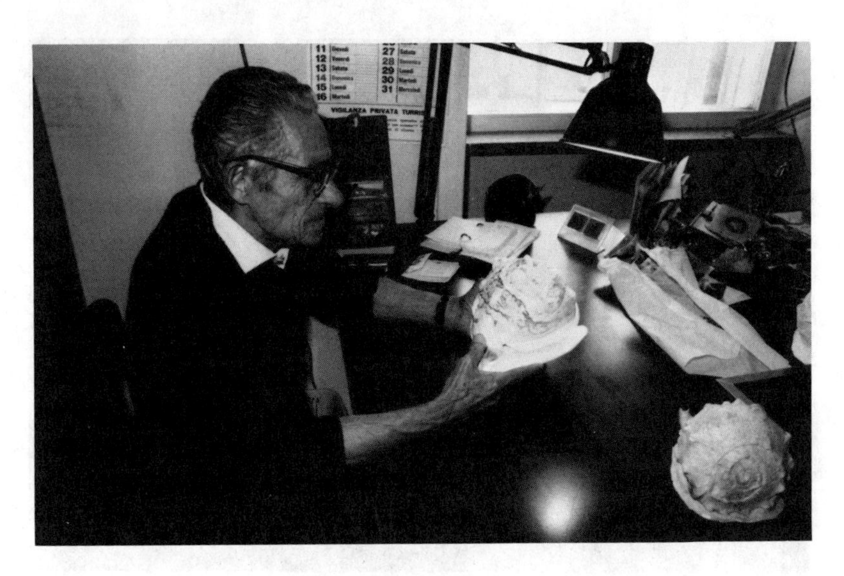

FIGURE 3-8.

*Dante Sebastianelli, an award-winning master carver, selects the perfect shell for a custom-ordered cameo.*

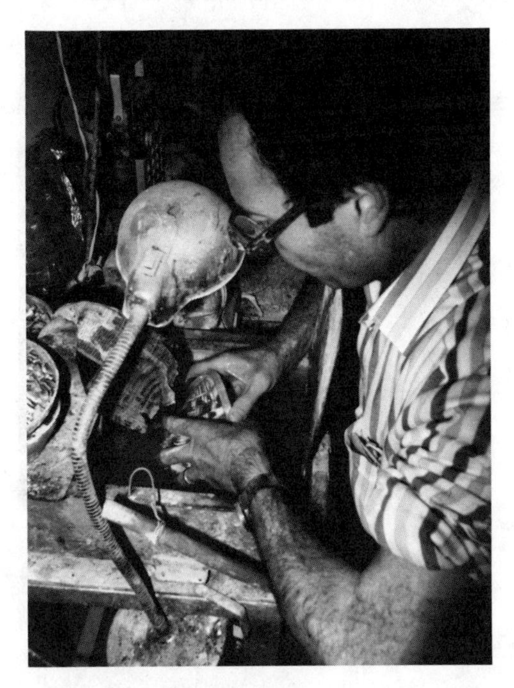

FIGURE 3-9.

*The selected shell is taken to the saw for preparation.*

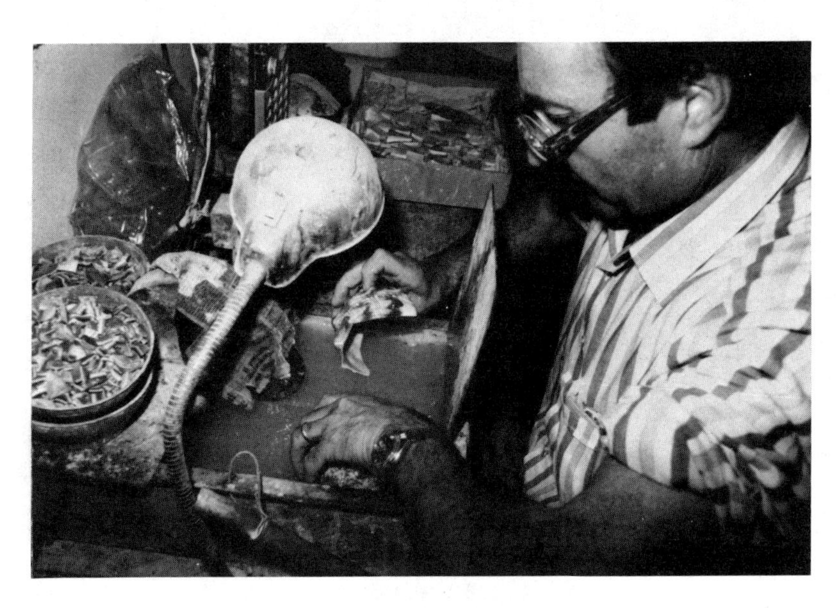

FIGURE 3-10.

*The shell is cut into two predetermined parts, the more important being the cup.*

FIGURE 3-11.

*Tourist souvenirs are often made from the part of the shell not used for fine cameos.*

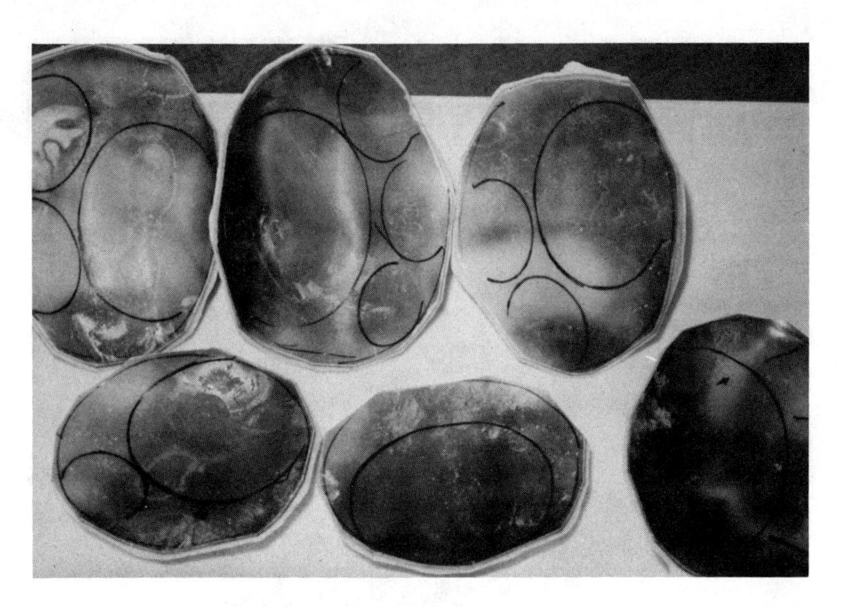

FIGURE 3-12.

*This is the cup, marked into cameo shapes.*

Marking the cup into shapes is called signing (fig. 3-13). The designer signs the cup with the shapes and sizes in indelible ink; the various sized outlines become the blanks from which the individual cameo is carved.

In the next step, the cutter takes over again and the cup is cut into sections. Shell that is too thin and other pieces that do not meet the requirements of the craftsman are discarded. The remaining pieces are sorted into quality grades and size groups (fig. 3-14). The blanks are then further reduced to calibrated sizes, and the rough edges are smoothed with a silicon Carborundum lapping wheel (fig. 3-15). The cutter's job is finished. The rough shell has been transformed into bicolored shell blanks of various sizes, ready for carving. The blank shells still have their original surface characteristics: the inner layer is rounded and smooth; the rough, curvilinear outer layer is later carved with the design chosen by the artist. Because the outer layer has depressions or protuberances, electric etchers or engraving machines are unlikely to be used, because they require materials with layers of even thickness.

In the last stage of the process, the piece of shell is attached by pitch to a round, wooden stick for easier handling of the material (color plates 40 and 41).

FIGURE 3-13.

*Marking the cup is called* signing.

FIGURE 3-14.

*After the blanks are cut from the cups, they are sorted by size.*

FIGURE 3-15.

*Calibrating sizes and smoothing rough edges of the shell blanks completes the cutter's job. The blank is now ready for the carver.*

Coral and shell cameo carving requires no diamond points, wheels, or discs. Several hundred years ago, the Italians were introduced to a carving tool called a *bulino,* a gentle scraping tool, the invention of the Jewish artisan Antonio Ciminiello. While in the twentieth century the dental drill is used for grinding away surplus shell, most carvers still use the same type of tools their forefathers had over two hundred years ago: steel gravers in numerous shapes and sizes (fig. 3-16). Traditional cameo carvers use three types of gravers: flat-faced, round, and three-cornered. Sometimes a double-edged knife is used as well.

The carver begins by sketching a scene or figures on the lighter parts of the shell. The design can be his own creation, or he may copy all or parts of a famous painting, such as *Birth of Venus* (fig. 3-17) or *La Primavera* (fig. 3-18). Or he may choose one of the dozens of myths, legends, historical subjects, or events that he finds interesting or has been commissioned to carve (fig. 3-19 and 3-20).

Carvers sketch at wooden benches near windows to take advantage of natural daylight. The sketch is given form by a V-shaped, thick, wooden-handled graver, which the carver uses to peck off the shell, a bit at a time. After roughly defining the design's outline (fig. 3-21), the carver

FIGURE 3-16.

*A traditional shell cameo graver with a cameo in progress on a dopstick, a blank awaiting sketching, and a finished piece.*

FIGURE 3-17.

*A shell cameo rendition of* Birth of Venus. *(Ann Spector Collection)*

FIGURE 3-18.

*A shell carver's interpretation of* La Primavera. *(Ann Spector Collection)*

FIGURE 3-19.

*A Garofalo shell cameo of three bacchantes. (Courtesy of Gennaro Borriello s.r.l.)*

FIGURE 3-20.

*Highly stylized shell carving of two Flora from the Sebastianelli company.*

FIGURE 3-21.

*Shell cameos in progress.*

FIGURE 3-22.

*Classic female profiles on shell cameos.*

changes to a fine-point burin and proceeds to cut the precise details of the cameo. Though a carver works quickly, depending upon the intricacy of the subject, a carving can take from a few hours to days to complete. The carver may use a hand-held, flexible-shaft electric grinding drill to grind away the white portion around the borders, but the majority of the work is done without electric tools. The newest rage among the Italian carvers is to use as much of the white layer as possible in the bas-relief. Carving it thinly to produce a shading effect, they sculpt it like marble, leaving only a small amount of the orange or brown background color as contrast (color plates 42 and 43). A shader—a small, square-headed tool—is used to smooth the background of the cameo. When it is finished, the cameo is soaked in olive oil, then washed in soap and water and polished with a hand brush. Much of the polishing is done on the background material only, to avoid abrading the bas-relief carving.

Most cameos produced in the last century depict classic, stylized female profiles (fig. 3-22). The trend among contemporary cameo artists is to produce the image of a more modern woman in profile, in scenes that convey movement: windblown hair, crashing waves, birds fluttering, drapery and garments moved gently by the breeze, dancers in motion, and so

FIGURE 3-23.

*A contemporary shell cameo from Sebastianelli.*

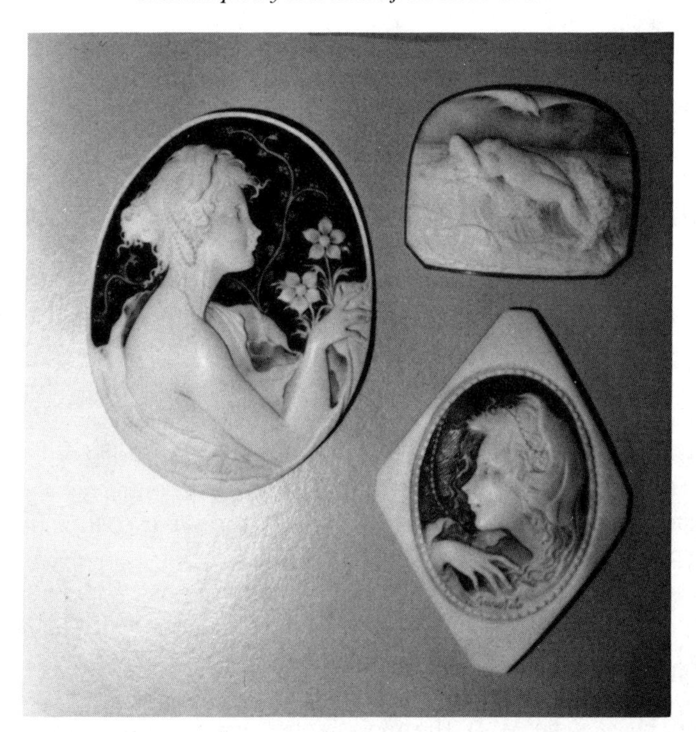

FIGURE 3-24.

*Cameos from the firm of Gennaro Borriello s.r.l. have modern shapes and subjects as well as classical ones. (Courtesy of Gennaro Borriello s.r.l.)*

forth. Static profiles and stationary scenes have been relegated to the past as today's carvers use innovative design to distinguish their work (fig. 3-23 and 3-24).

---

## CORAL

Coral presents a certain mystery: scientists cannot explain why some portions of the seabed are covered with coral and others, seemingly identical, are not. Nor can they tell how long it takes a coral reef to reach maturity or why some corals are red while others are pink, white, or shaded. The nature of coral has remained a puzzle for thousands of years. At first coral was pronounced a mineral; then in 1706 the Italian naturalist Luigi Marsili called it a vegetable; twenty years later, a french scientist named Peyssonel declared it an animal. Though classified as an animal today, it remains an enigma.

Regardless of its nature, coral has been a symbol of beauty, hope, and protection against evil since classical times. Many European archaeological and ethnographical museums display coral charms and ornaments dated as far back as the first millenium B.C. Coral fragments that appear to have been amulets have been found in neolithic-period graves. The ancient Greeks and other Mediterranean cultures believed coral to be useful against the "evil eye." The connotation of protection continued into the Christian era, when coral was made into crosses and charms carved with the figure of a lamb, to be worn on the neck or wrist. Coral is still associated with suffering and is used today in some cultures to adorn children and to help them cut their teeth. Many old-master paintings show children in repose or at play with strands of coral beads around their necks. In the Victorian era, babies were given coral pacifiers to ease the pain of cutting teeth, as well as to protect them from falls and sickness, lightning, fire, fits, sorcery, and the evil eye. Parures of coral beads and carved coral were fashionable wedding gifts throughout the nineteenth century.

Like many hardstones, coral was given gender: red denoted the male, and white the female. Coral has long been used in folk medicine, its powers presumably ranging from curing specific ailments to use as an aphrodisiac.

In the fifteenth century, the Sicilians transformed coral into a great art form; the workers of Trapani produced the most magnificent coral jewelry of the times. Coral also decorated the clothes of the nobles, clergy, and the fashionable. Coral carvings were sought after to fill the great collections of royalty. Branches of coral, polished and mounted in gold, carved coral necklaces, and other coral carvings are on view today in the

Topkapi Palace museum in Istanbul, Turkey, part of the collection of abundant riches once owned by the Ottoman sultans.

At the start of the nineteenth century, coral art expanded considerably, particularly in Torre del Greco, Italy. The discovery of new coral beds around Sicily caused a factory especially for coral art to be built in Torre del Greco. The city became the chief European center for the production of coral items, as well as the most important port for coral fishing. Today most coral fishing is done by divers combing the Sardinian and Spanish seas and the waters around Sicily and North Africa. Other coral comes from the seas near Japan and China. If coral is from the Mediterranean region, it is identified as *noble;* it if is from any other source, it is referred to as *Japanese* coral.

The noble coral of the Mediterranean is found in 20- to 25-inch-high shrubs, weighing little more than a pound each. The slender branches are regularly spread, sometimes even bushlike. It is found in a variety of colors, and the coral carvers make a fine distinction among coral tints: pure white is *bianco;* pinkish-white is *pelle de angelo;* pale rose, *rosa pallido;* bright rose, *rosa vivo;* red, *rosso;* dark cherry red, *rosso scuro,* and darkest of the red tones, *carbonetto* or *ariscuro.*

Though it is smaller than Japanese coral, Mediterranean coral is more valuable because of its harder and more compact grain, which provides a better carving surface. The price of raw coral is determined by weight as well as color.

Coral is a product of sea life, and its composition is mainly calcium carbonate. It is a soft material, easily worked with a file and edged tools; the common method being by hand, under running water, to prevent cracking. The finished pieces are polished with linen and burlap. Carvers of coral must be well acquainted with the material because the waste is great, sometimes as much as 90 percent, which accounts in part for the high cost of a fine coral cameo.

Any accounts of ancient carvers making coral cameos or seals is missing from the history of the glyptic arts, and scholars generally agree that the prolific use of coral, either for relief or intaglio work, began during the Renaissance.

Coral cameos of the last century are often found in a wide range of orange-red colors, especially in the stock of dealers of antique jewelry (figs. 3-25 and 3-26). Most are of poor to mediocre quality. Finding a truly artistic, well-preserved antique coral cameo of a desirable color, such as that in figure 3-27, is an exciting challenge to the collector.

Japanese coral grows much larger than that in the Mediterannean, as much as 3 feet in length, weighing four to five times more. It has big

FIGURE 3-25.

*Orange and red coral cameos from the nineteenth and twentieth centuries.*

FIGURE 3-26.

*A bright red coral cameo of a Medusa head, from the mid-nineteenth century, set in a modern 18-karat-gold ring mounting.*

FIGURE 3-27.

*An angel-skin coral cameo of angels, circa 1870. (Photo courtesy of L. T. Hall, Windsor, VT)*

branches that protrude to the left and right, and a large root diameter, from ¼ to ¾ inch and up, which provides perfect material for statues. Colors of Japanese coral range from dark red to cherry red to pink.

## CARVING CENTERS AND RENOWNED CARVERS

### IDAR-OBERSTEIN, GERMANY

Since the middle of the eleventh century, minerals have been mined around Idar-Oberstein, when an agate mine, the Steinkaulenberg, was opened near the city (fig. 3-28 and color plate 44). Records from 1497 show miners worked even then with hammers, pickaxes, and chisels to remove agate, jasper, and amethyst from the maze of underground passages.

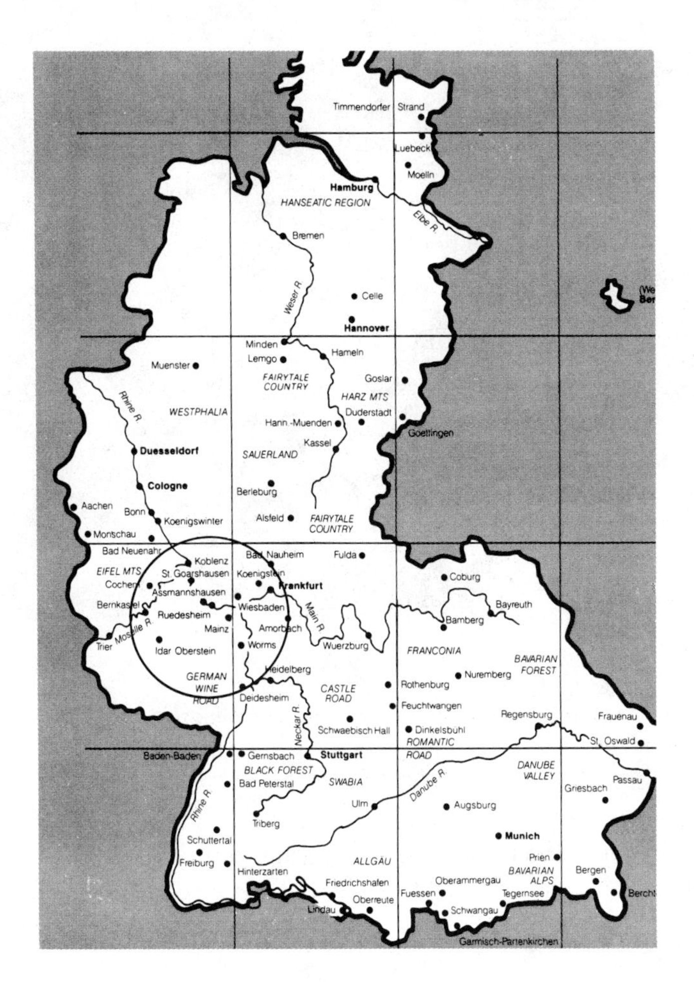


**FIGURE 3-28.**

*The German city of Idar-Oberstein, a pleasant two-hour car journey from Frankfurt, has been a center of gem cutting and engraving for over five hundred years.*

Some believe that gem cutting and engraving began in Idar as early as 50 B.C., when Julius Caesar visited the area. However, that is conjecture; what is known for certain is that the abundance of mineral material resulted in the building of countless grinding mills, powered by waterwheels driven by the Nahe River. The huge sandstone grinding wheels were operated by workmen lying on their stomachs on benches, pressing the rough stones against the massive grinding wheels (fig. 3-29 and color plate 45). This initial preparation gave the rough material form and luster. Grinding was a

FIGURE 3-29.

*The interior of a typical mill in Idar-Oberstein. The sandstone wheels, belts, and pulleys were powered by water.*

strictly regulated trade. A document from the 1609 statutes of the Grinders' Guild makes plain who could apprentice to that craft: "No stranger must acquire the trade of a grinder, but the craft should be handed down from father to son."

Following the Napoleonic wars, the area around Idar was in a state of siege and devastation. With the mineral supplies all but exhausted, many German carvers left the country to seek their fortunes elsewhere. The emigrants to Brazil soon discovered agates and other minerals that were just what was needed back home in Idar. The shipment of minerals from Brazil to Idar-Oberstein began in 1834 and continues to this day. Some of the emigrants returned to Idar with knowledge of new gems, and with the fresh influx of raw material, the mills were revived. In the late nineteenth century, more than three hundred water-powered grinding mills were in use along the Nahe River. With the advent of electricity, in about 1900, the grinders, no longer dependent upon water to turn their wheels, moved their workshops into their homes. Today it is rare to see a grinder work a sandstone grinding wheel. Most of the old wheels and mills are now valuable artifacts and have been either turned into museums or fallen into disuse.

The sandstone grinding wheel was never used to process stones harder than quartz; this was done with a horizontally spinning disc, a process introduced into the region about one hundred years ago, at the same time cameo and crest carving began. The German craftsmen became widely known for their rich ornamentation of vessels and their gem sculptures.

Presently in Idar there is a group of highly acclaimed engravers who create bowls, birds, and animals from a wide variety of minerals. Master bowl carver Helmut Wolf has a long list of honors to his credit, as does gemstone carver Bernd Munsteiner, a carver who devised a gem cut called the "fantasy cut." This is his own uniquely developed method of cutting a gemstone to produce a sculptural effect. Other well-known carvers (past and present), gem cutters, and companies in Idar-Oberstein include: Edwin Pauly, Hans Pauly, Richard H. Hahn, Karl Hahn, Gebruder Leyser, August R. Wild, Johann Philipp Wild, Georg Wild, A. Ruppenthal, Friedrich A. Becker, Gerhard Becker, W. Constantine Wild, Otto and Dieter Jerusalem, R. Jachem, Paul Dreher, Max Herringer, Kurt Becker, Helmut Bussmer, Aray Dommer, Werner Kohler, Heinz Postler, Jurgen Reinhard, Hans Dieter Roth, Jurgen Thom, and Hans Waldmann. These are but a few of the better-known craftsmen in the Idar-Oberstein gem industry; countless others have worked in both master and apprentice positions.

Engravers all over the world admit that nowhere except in Idar-Oberstein can one learn the craft of hardstone engraving and cameo carving using traditional centuries-old methods. And only there is gem carving taught formally. The course of study includes art and art history, gem technology, business management, and training in basic art skills, such as anatomical studies. In 1990 there were approximately two hundred cameo carvers in the Idar region.

There are some specific educational requirements for carvers that parallel those of the sculptor: drawing from life and the ability to replicate the human face in clay; the ability to create a model and depict human anatomy. Most carvers must practice drawing to be able to sketch their subjects on stone. An early transition step between sculptor and carver is the ability to model bas-reliefs in wax. The technique of first modeling a subject in wax, then transferring it to stone, was used by many leading cameo carvers of the eighteenth century. It was a quick way to fix an image before the advent of photography. A Roman, Benedetto Pistrucci, was able to make wax models so quickly and efficiently that he was repeatedly commissioned by the English court. One of his most famous pieces that later became a cameo was a wax model of Queen Victoria for her coronation.

---

## Torre del Greco, Italy

Nestled at the foot of Mount Vesuvius on the Bay of Naples is the ancient town of Torre del Greco (fig. 3-30 and color plate 46). This picturesque Italian fishing village has a worldwide monopoly on shell and coral cameo production, which they mantain with a labor force of about five thousand artisans. (Basilio Liverino, a third-generation head of a prestigious Torre del Greco coral firm bearing the family name, has one of the world's finest museums of carved coral, objets d'art, and jewelry.) They also carve mother-of-pearl, tortoiseshell, and ivory (declining with the worldwide ban) and manufacture gold and silver mountings. Cameo carving is an ancient skill here, where the ruins of Pompeii, only a few kilometers away, have yielded stone cameos preserved after the city was destroyed by a volcanic eruption in A.D. 79.

Coral predated shell as carving material, but documents allude to both materials being worked here in the eleventh century. The first real

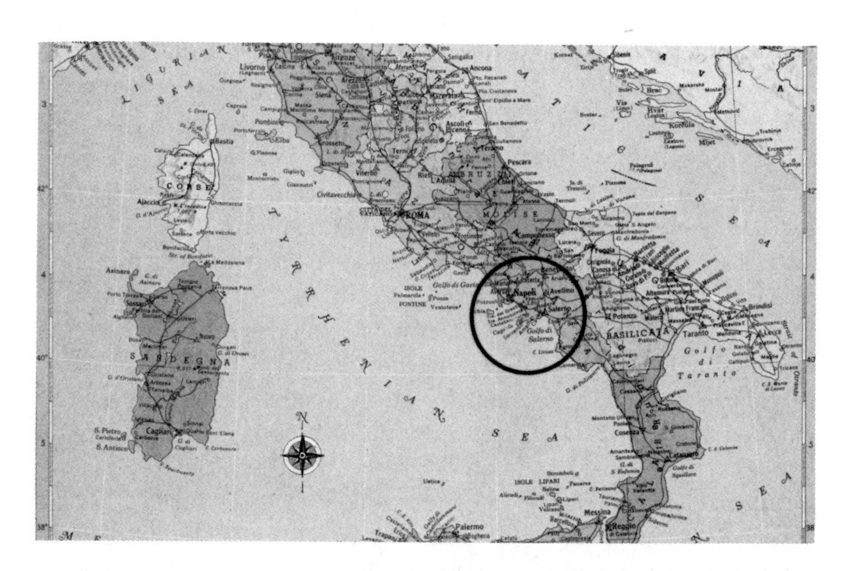


**Figure 3-30.**

*The Italian town of Torre del Greco is a few miles south of Naples.*

workshops were set up much later, however. In the eighteenth century, the fishermen monopolized coral fishing and distribution, and from their ranks emerged a new class of artists and engravers. These men were graduates of the Royal Workshop of Hard Stone of Naples that was opened to train artists and craftsmen in coral, shell, and lava production. The supremacy of the Torrese artisans was so great that on June 23, 1878, the School for the Manufacture of Coral and Cameos ws founded by a royal ordinance of Umberto I in the same town. From this time forward, Torre del Greco and cameo carving have been intertwined. Today the craft is taught at the town's Institute of Fine Arts, and the products are publicized worldwide. Some of the world's finest craftsmen are graduated from the school, and some individual carvers have developed international reputations. Torre del Greco is the only town in the world where shell and coral carving is formally taught.

There are several leading cameo producers in Torre del Greco, among them, Sandro Sebastianelli, Giovanni Apa, Gennaro Borriello s.r.l., and Basilio Liverino. All have their own workmasters; a few work off the factory premises, designing and carving in their own ateliers.

Sandro Sebastianelli is head of the Sebastianelli firm in Torre del Greco, now in its third generation of shell cameo production. Sebastianelli believes that people are once again beginning to understand the art and craftsmanship involved in making a fine cameo. "A cameo is a piece of art," he says, "but until the last few years, the art has not been understood in the United States. The entire world market is beginning to appreciate quality at last. The post-World War II period was the worst time in this century for cameo production." In the United States, most cameos that are sold have a brownish background, Sebastianelli points out, theorizing it may be because the carnelian cameos were the first his firm shipped to the United States. Current fashion has sparked an interest in cameos, but he does not expect the rise in popularity to be only a trend. Instead, he thinks consumer interest will continue to rise and cameos will become a jewelry staple, like pearls. Sebastianelli believes that contemporary cameos are affordable art forms, with prices ranging from $50 to $400 for an average cameo; some original art sells in the $1,000 to $3,000 range, without a mounting.

Dante Sebastianelli is the patriarch of the company, himself an award-winning carver in his younger days. One exquisite work evolved from a dream; he rose from his bed and set to work immediately on the design, not stopping until it was finished twelve hours later (fig. 3-31). The elder Sebastianelli requires that his master carvers put their heart and soul into a design and feels that the most critical test of a carver's talent lies in his

FIGURE 3-31.

*The subject of Dante Sebastianelli's award-winning cameo came to him in a dream. The cameo depicts many stages in a man's life.*

ability to make the subject live. In fact, he puts it in very simple terms: "If the flowers have no smell, the carved image is poor work."

A master carver must have more than just talent for design. He has to study the thicknesses and colors of the shells before he chooses a subject, whether it is a portrait, mythological scene, or reproduction of a masterpiece. Whatever the choice of design, it must be executed within the thickness of 1 or 2 millimeters while taking into account the depth of the perspective, the shading, and every possible effect produced by the transparency of the layers.

The company of Borriello s.r.l. was founded in Torre del Greco in 1954 by Gennaro Borriello. From the beginning it was a company that manufactured coral and shell cameos. Today the company not only imports the shells and designs and produces the cameos, but also follows through with the production of mountings to make finished pieces of jewelry.

In the beginning Borriello was a shell carver, and his wife, Marianna Belardo Borriello, was a coral artist. The couple have taught their own artisans, and today the family business employs about fifty craftsmen. A daughter, Antonella Borriello, is vice-president of the firm and oversees the day-to-day production operations. Several internationally famous carvers work for the firm: Genaro Garofalo, Francesco Esposito, and Salvatore Balzano. Their cameos are signed and displayed in museums all over the world.

The Borriellos emphasize that the shells they use are not dyed but naturally, uniformly colored; only all-natural colored shells are used. They believe that only the male helmet shell *(Cassis tuberosa)* has the thick layers required for a fine carving; female helmet shells have thin layers and are discounted as good carving material.

Antonella Borriello says that while some of their carvers have attended art school, many have not. A certificate or diploma in art is not required to get a job as a carver; neither is membership in a guild or association. What may be necessary, however, is a lengthy apprenticeship period, even if the man shows artistic ability. It can take up to five years to learn how to start and finish a cameo properly; once learned, the skill is maintained throughout the artisan's working career, as vocation changes are rare. At the present time Borriello employs no women as either master or apprentice cameo carvers.

The subjects of the cameos are decided by owner Gennaro Borriello himself. A carver may work on a single cameo without interruption, from start to finish, or he may work on several different subjects at one time, alternating among them as one becomes tedious. Some artists find working whole helmet shells to their liking because of the multiple figures and larger work space (fig. 3-32). Some of the carved whole helmet shells not only have scenes of arresting beauty, but are historic and world famous. The Neapolitan sculptor Giovanni Sabbato spent ten years carving what has become the best-known helmet shell cameo in the world, *The British Empire,* circa 1873, which is about 7½ by 6½ inches, or 20 by 17 centimeters (color plate 47). This masterpiece shows Queen Victoria with allegorical figures of the empire; national, civic, and public figures; symbols, seals, crests, banners, and standards. It can be seen today in the Basilio Liverino Museum in Torre del Greco.

When asked to separate the old shell cameo carvings from the new, Antonella Borriello gives the following hints: new shell cameos have thick, white borders; women's faces and bodies show modern interpretation; men seldom appear in the newer cameos. Some carvers leave a subtle hint of their artistry—almost a signature—by the way in which they carve certain parts of the body, such as hands. The faces and bodies of women featured in the cameo carvings of Genaro Garofalo and Salvatore Balzano are distinctive, highly stylized individual presentations; they look like contemporary women. Because so many commercial cameos are on the market,

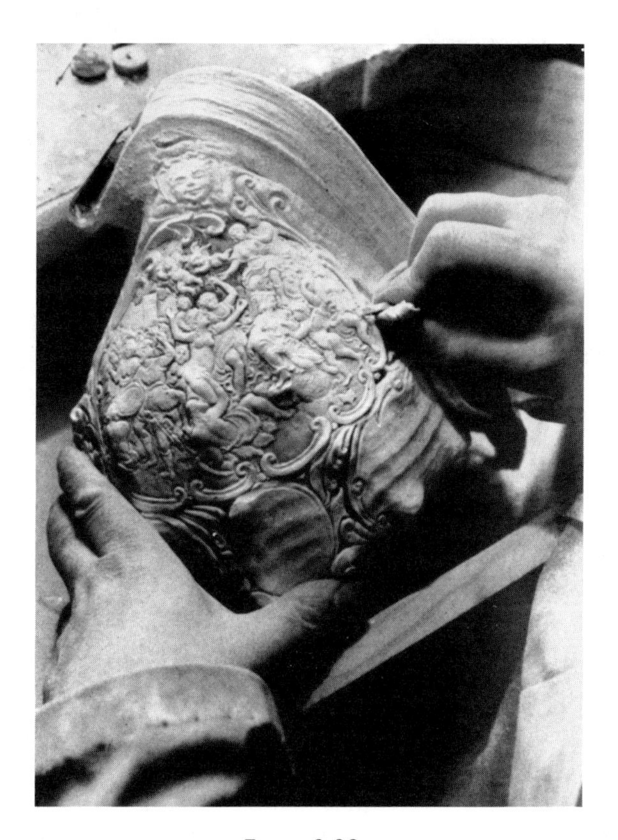

FIGURE 3-32.

*Carving the entire shell as a cameo enables the carver to lavish his artistic talents on the piece.*

the few artists who sign their work are not as well known as some of the German hardstone carvers, even though their artistic talents may be equal.

Companies that are engaged in mass production have carvers that can furnish up to twenty pieces of one style a day. A kind of production line is set up, with one man carving and another doing finishing work or polish and details. This kind of work is never signed, and the carvers are anonymous.

## JAPANESE CARVERS

Masashi Furuya and Company, in Yamanashi-ken, Japan, is a cameo-carving factory that only a few years ago supplied 100 percent of the country's market for cameos, produced in shell, hardstone, and coral. According to the executive manager, Masashi Furuya, Japan is the largest market in the world for cameos, so it was quite natural that most of the Japanese production was sold there. Today, the Japanese also export their cameo art to the United States, Canada, and Europe.

Cameo carving was introduced into Japan by Furuya in the early 1970s. During a business stay in Idar-Oberstein, he saw the hardstone cameos of the German masters and began to learn about and enjoy glyptic arts. After studying how carvers used the onyx stone layers to their best advantage to produce cameos of universal aesthetic appeal and salability, he put this knowledge to work back in Japan. Furuya commissioned cameos from a young German carver, Gerhard Schmidt, who had fifteen years of study and was a former pupil of Richard H. Hahn. Schmidt visited Japan in 1981 at the invitation of Furuya to demonstrate cameo carving and give technical assistance to Japanese artisans interested in learning the cameo art. Consequently, Schmidt took part in the first private cameo exhibition in Japan, in June 1982, at the Yamanashi Prefectural Art Museum.

Today the Furuya factory imports agate from Brazil to make both hand-carved and ultrasonically carved cameos. Cameos are produced for pendants, rings, earrings, and brooches, which they export. Because of the European training in design and technique, there is no simple way to distinguish between the agate cameos that come from Japan and those from Germany.

## THE FUTURE FOR CAMEO CARVERS IN THE UNITED STATES

The future for cameo carvers in this country has never looked better. There have always been a few artisans in the New York area, such as Ottavio Negri and Beth Benton Sutherland; but in 1960 a trained cameo carver from Idar-Oberstein, Ute Klein Bernhardt, arrived in Chicago to live and work. Bernhardt, a former apprentice to Richard H. Hahn, had polished skills and enviable credentials. Among her notable commissions was a cameo portrait of Marshall Field I, carved in 1,800-carat Russian beryl, displayed at the Field Museum of Natural History. She is instructing a new generation of young talent in cameo carving and in so doing has been instrumental in bringing new vitality to the art. The future for the cameo looks bright: with new artists to glorify it and new patrons to encourage their work, new collectors are certain to find and appreciate it.

## INSTANT EXPERT

### *Common Materials for Engraved and Carved Stones*

Neither the Greeks nor the Romans had any scientific knowledge of gemstones, and more than one kind of stone was often used and classified under the same name; for example *adamas* is used for both diamond and white sapphire.

| | |
|---|---|
| Egypt (3100-1000 B.C.) | Amethyst, glass, lapis lazuli, turquoise, green feldspar, green and black basalt, jasper, carnelian, crystal, soapstone, alabaster; amethyst frequently used for scarabs |
| Mesopotamia (2000 B.C.) | Serpentine, marble |
| Assyria (1400 B.C.) | Rock crystal, agate, carnelian, amethyst in cylinder seals |
| Greece (700-100 B.C.) | Cryptocrystalline quartz, garnet, beryl, adamas, glass paste, amber, ruby, sapphire, sardonyx, soapstone; amethyst used, but infrequently |
| Romen (100 B.C.-A.D. 500) | Sardonyx, agate, jasper, garnet, aquamarine, lapis lazuli, ruby, sapphire, rock crystal, glass paste, adamas, amber, bloodstone, topaz, peridot, turquoise; prodigious use of amethyst |
| Renaissance (1400-1600) | Shell, coral, mother-of-pearl, ivory, agate, onyx, garnet, lapis lazuli, carnelian, ruby, glass paste |
| Georgian (1698-1830) | Agate, onyx, shell, coral, malachite, emerald, glass paste |
| Victorian (1837-1901) | Shell, multilayered onyx, lava (green-gray, beige), coral, agate, chalcedony, malachite, sapphire, garnet, jade, opal, quartz, jet, glass paste |
| Modern (1901- ) | Agate, onyx, shell, turquoise, spinel, ruby, sapphire, diamond, lapis lazuli, garnet, quartz, aquamarine, emerald, opal, peridot, topaz, tourmaline, zircon, zoisite, glass, plastic, synthetic stones |

Chapter

4

# DISTINGUISHING OLD AND NEW CAMEOS

There is no scientifically acceptable way to establish the date a cameo was carved. To attempt dating by examining the style of a cameo's setting or mounting is folly, because cameos are so easily remounted to suit their owners.

Some eminent nineteenth- and early-twentieth-century authors on the subject have given hints for fixing the approximate carving date of cameos, but most of the clues are found in out-of-print French, German, or Italian books, and most of the information is faulty in any event.

Differentiating old cameos from new depends upon acquired knowledge of designs, inscriptions, and carving styles that were common in the past. The primary element necessary for differentiation is personal knowledge of the glyptic arts. Until space-age technology can provide some exotic scientific method, circa-dating expertise is still gained through study.

Derek Content, of Houlton, Maine, is an expert on engraved gems and cameos who recently had an exhibition of his personal glyptic-arts collection at the Ashmolean Museum at Oxford University in Great Britain. When asked for a formula to help distinguish old cameos from new, he stated flatly: "Spend twenty years looking at them. There were good and bad carvers in every period, but they all carved in their own milieu. The gems, therefore, are of a certain style of the period, except for the deliberate forgeries, or copies from another period. If you want to know how to tell them apart . . . well, jewelry historians have been arguing about that subject for the past three hundred years. It comes down to getting a basic feeling for the material, the styles of carving, and the epochs in which they were carved. There are no shortcuts."

Content punctuated his advice with the story of a boy wishing to become a jade carver: "A young boy wanted to carve jade the way his great-grandfather, grandfather, and father did. So, the grandfather, father, and boy all sat around in a circle, handing rocks around to touch and feel. The ritual was carried out for hours each day for several months, until finally in exasperation the boy cried: 'All we ever do is *feel* these stones, and I want to begin carving! In fact, this stone you have just handed to me doesn't even *feel* like jade!' 'Ah-ha!' exclaimed the grandfather, 'at last you are beginning to learn about jade.' "

The lesson is clear: to learn about any gemstone material, engravings, or cameo carvings, one must handle the materials daily and develop a sensitivity for them. For most buyers and collectors of cameos, this means education primarily from observation, at the many museum exhibitions, the private collections that are open to the public (usually at gem and mineral shows), the dealers specializing in cameos, and the auction houses dealing in estate jewelry, which often includes cameos. Buyers of cameos, collectors, and jewelry appraisers are advised to acquire specific and in-depth knowledge of the subject before voicing any firm opinion of a cameo's origin.

## CIRCA-DATING CAMEOS

A cameo carved in garnet, lapis lazuli, or emerald may be dated in some cases. Some gemstones can be associated with particular periods of discovery and use, as well as specific mining sources, because of their distinctive inclusions. This kind of determination, however, is best left to a graduate gemologist with a well-outfitted laboratory. With proper research and analysis, the gemologist may be able to establish the country of origin of the gemstone and the period of its most popular use. This is a very generalized kind of provenance, questionable at best. Better attempts to uncover origin lie in the study of the technique and virtuosity of the carver, along with investigation of the styles and history of the period in question. Much can be gained by having a thorough understanding of ancient cultures, because through the universal language of art, the carver—though long dead—has left a clear and vital message.

Portraits and inscriptions that name rulers are not very accurate guides to circa dating, because some rulers were venerated long after their deaths. Some, however, were despised after their reigns, so if either their names or portraits are found on cameos they should be considered contemporary to the ruler's reign.

Portraits found on ancient coins can often be matched with portraits on cameos. This is only an indicator of a possible date. A face on an ancient coin is of more value as a means to identifying the portrait on a cameo than to dating it.

My book *Illustrated Guide to Jewelry Appraising: Antique, Period, and Modern* points out how the noses on profiles have undergone a dramatic change over the years. The classic Greek and Roman faces have straight noses (figs. 4-1 and 4-2), whereas the noses of women in a later period show a slight tilt upward. Those carved in the twentieth century look quite perky and adorable, with upturned noses that make the subject a lookalike for the modern-day magazine covergirl (figs. 4-3, 4-4, and 4-5).

The drawing of a woman's profile is also a clue to dating, according to Rose Goldemberg in *Antique Jewelry: A Practical and Passionate Guide.* She noted that simple, classical heads of women and men are usually seen in early cameos; whereas the fussier, Victorian ones from 1850 onward depict women with heavy necks, jowls, and matronly airs. This is especially true of the mass-produced late nineteenth-century cameos.

At the turn of the century and into the 1900s, cameos with sensu-

FIGURE 4-1.

*An agate cameo of a bacchante with a vine wreath in her hair and a ram's head on her shoulder. This neoclassic cameo has been dated to* A.D. *1700 to 1850. (The Sommerville collection, University of Pennsylvania)*

FIGURE 4-2.

*A neoclassic sardonyx bacchante. (The Sommerville collection, University of Pennsylvania)*

FIGURE 4-3.

*A finely carved hardstone cameo by an early-twentieth-century German master carver. Noses on antique cameos are usually carved in a straight, Roman style. The more contemporary carvings depict different shapes. (Ann Spector Collection)*

FIGURE 4-4.

*This Sebastianelli cameo depicts modern women with upturned noses.*

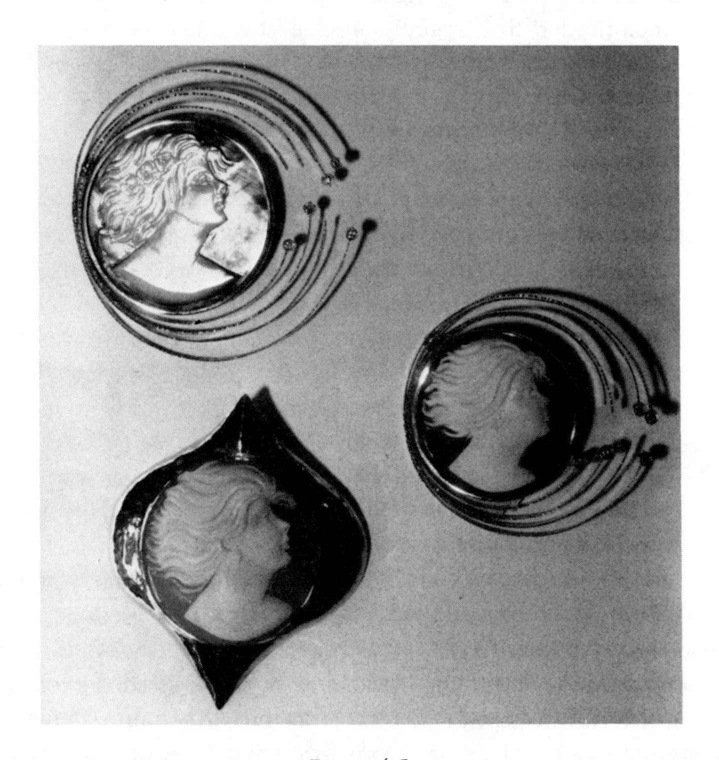

FIGURE 4-5.

*These are Sebastianelli profiles of modern women with short, wavy hair and small noses.*

ous art nouveau sirens and images of the new, daring, modern woman were carved (color plate 48). Some cameos of the 1920s are characterized by the woman's short, bobbed hairdo, even to the neat fingerwaves carved on their heads. The women's social and political movement promoted portraits of great athletes, such as tennis champion Helen Wills Moody (color plate 49), as well as celebrated stage personalities.

---

## Anatomy

Modern-day carvers do not interpret the human body in the same way as their predecessors did. Muscle tone is articulated differently in Hellenistic, Renaissance, nineteenth-century and twentieth-century cameos; The Greeks, for example, were masters at carving animal and human figures. Researchers point out that all the great Greek engravers used mythological and heroic subjects instead of subjects contemporary to their time. Engraved-gem expert and writer John Boardman has noted that ancient Greek artists lavished their best skills on engraved gems and used a cable pattern around their borders that is purely Greek in style and execution. He also said that anatomy in the sixth century B.C. has a certain fullness or fleshiness not seen in work from other periods or cultures.

---

## Hairstyles

The peculiar hairstyles of men and women found in the archaic and ancient cameos may help circa-date them. Men wore short hair and followed the Macedonian fashion of shaving up until the second century A.D. A hairstyle for men called the *coif* was fashionable for most of the thirteenth century. Many drawings of the style, popular in France beginning in 1235, prove its popularity. Historians tell us it was worn into the fourteenth century, only to completely disappear by 1340. The coif can best be described as a close-fitting hairstyle, almost like having a cap on the head, with fringes of hair uncovered over the forehead and at the nape of the neck.

In the sixth and seventh centuries B.C., Greek women wore their hair long, commonly falling over the shoulders and down the back, usually fastened by a headband or diadem. In about 500 B.C., the hair began to be restrained, with the ends either tucked up under the headband or gathered into a net on the nape of the neck. From the fifth to fourth centuries, it was more common for hair to be worn up in a bun. Many vases and articles of sculpture are witness to this style. In the Hellenistic period hair was curled into a variety of styles.

The portraits in figures 4-6, 4-7, 4-8, and 4-9 are typical women's hairstyles in the times indicated. These same styles can be seen on statues and bas-reliefs on monuments in museums all over the United States, Europe, and the Middle East, as well as in cameos in numerous university collections (figs. 4-10, 4-11, and 4-12). The danger in using these coiffures to circa-date cameos is that they are useful for identification of the subject only and may not necessarily correspond to the actual date of the cameo carving (fig. 4-13).

Carvers in the sixteenth century, while excellent craftsmen, were not very creative. They copied and replicated classical Greek and Roman subjects from design books that were widely circulated and used. Therefore, dating a cameo on the strength of the hairdo alone is not wise. Individual clues must be considered in the context of the entire piece. Notwithstanding all the above, hairstyles are an interesting indicator of society and

FIGURE 4-6.

*Hairstyle of the first century A.D. (Illustration by Elizabeth Hutchinson)*

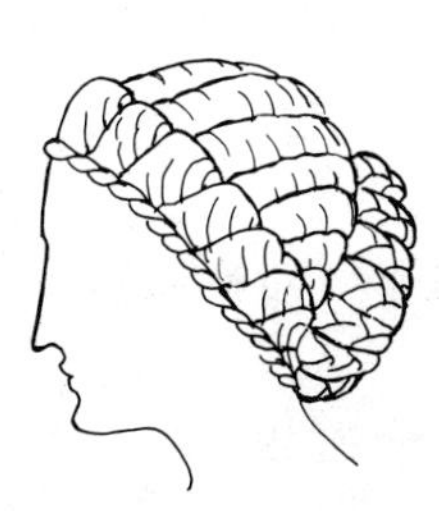

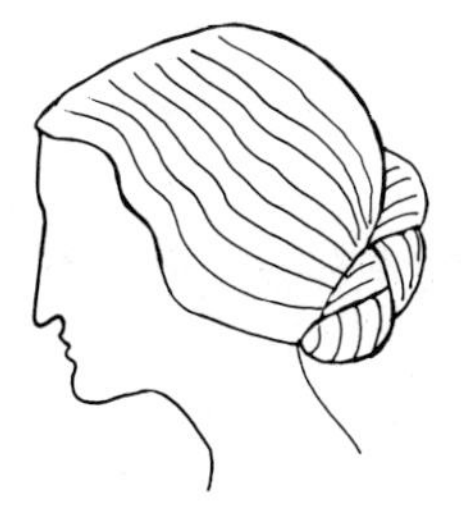

FIGURE 4-7.

*Crispina-type hairstyles, from the second century A.D. (Illustration by Elizabeth Hutchinson)*

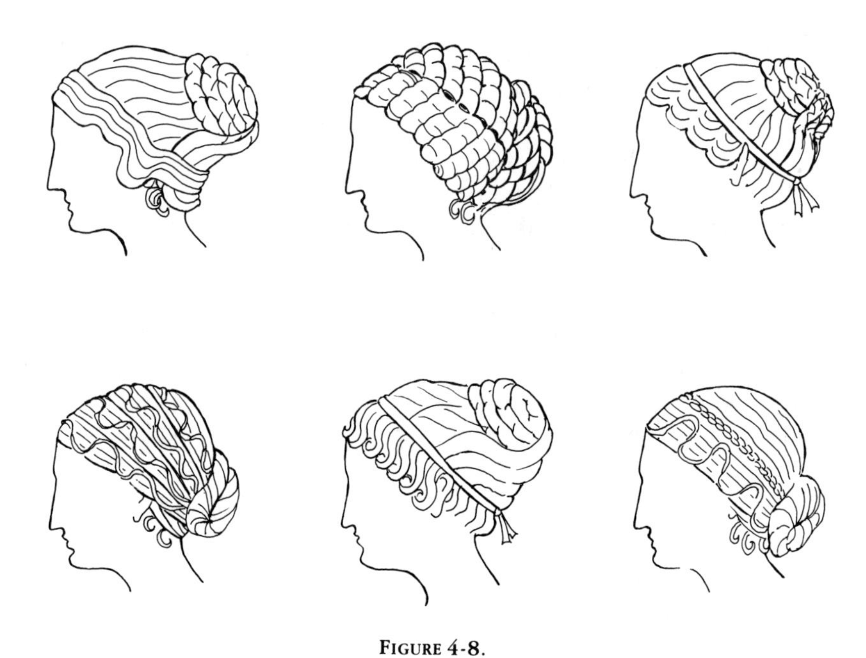

FIGURE 4-8.

*Faustina-type hairstyles, from the second century A.D. (Illustration by Elizabeth Hutchinson)*

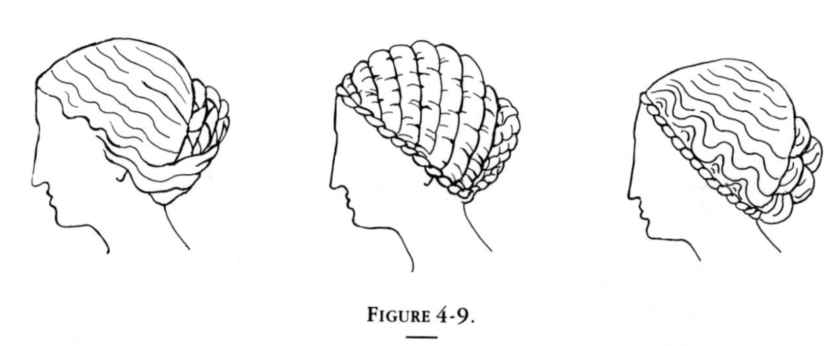

FIGURE 4-9.

*Lucilla-type hairstyles, from the late second and early third centuries A.D. (Illustration by Elizabeth Hutchinson)*

FIGURE 4-10.

*This female profile in onyx is from the second century A.D. and has a hairstyle identified as a Crispina type. The cameo measures about 1½ by ⅞ by ½ inch (40 by 23 by 13.8 millimeters). Ex Harari collection. (Content Family Collection of Ancient Cameos, Houlton, ME)*

FIGURE 4-11.

*A female portrait with a Faustina-type hairstyle, carved in sardonyx.*

FIGURE 4-12.

*A sardonyx cameo of a woman with a Lucilla-type hairstyle, from the second or third century A.D. (The Sommerville collection, University of Pennsylvania)*

FIGURE 4-13.

*Even though this beautifully carved cameo has a Grecian portrait with a hairstyle that can be identified as a Faustina type from the second century A.D., it is not an ancient cameo. The nineteenth-century cameo is a reproduction of a classical style. (Photo courtesy of Achim Grimm)*

FIGURE 4-14.

*Contemporary cameos generally show women with long hair or masses of hair. (Photo courtesy of Achim Grimm)*

fashion trends, from the long curls and romantic ringlet coiffure styles of the Victorian period (color plate 50) to the short hairdos of the 1920s and unrestricted, flowing mane of more recent years (fig. 4-14). Of one thing we can be certain, however: if a cameo is carved with a female wearing a hairstyle from the Victorian or later period, it obviously could not possibly be an archaic engraving, as early carvers could not articulate what they had never seen.

## JEWELRY

Female subjects of ancient cameos seldom were depicted wearing jewelry. Athena may be found with her owl, Diana with a quarter-moon, but hardly a female will be found with jewels in her ears or around her neck. Diadems, ribbons, nets, and wreaths for the head and hair, however, were popular. This lack of jewelry is surprising in view of the fact that jewelry was much prized by the women of classical antiquity, and it is found in coin portraits and some other art forms. Gifts of jewelry were mentioned in

FIGURE 4-15.

*This black and white onyx carving by a contemporary German master carver displays the almost obligatory addition of jewels to the portrait. (Ann Spector Collection)*

literature; in *The Odyssey,* for example, there is a description of gifts with which suitors tried to seduce Penelope: "Brooches, twelve in all, fitted with curved clasps; a chain, cunningly wrought of gold and strung with amber beads, bright as the sun; a pair of earrings with three clustering drops, and a great grace shone therefrom; a necklace, a jewel exceeding fair." The Greek women on vase paintings, on coins, and carved in statues were more likely to wear one or more trinkets or items of jewelry than those shown in cameos. An entire series of statues found on the Acropolis at Athens portrays women wearing jewelry: large disc earrings and possibly jeweled diadems. Diadems were used throughout classical times. Garlands were also worn from early Greek to late Roman times, generally conferred by the state as a mark of honor or worn in a religious procession. They were also employed for funereal purposes, placed on the head as a mark of victory in the battle of life. Some of the wreaths were of thin, beaten and pressed gold sheets formed to represent vines, somewhat like jewelry. The diadem, an ornamental headband, was set with large garnets or red stones, and there were loops at the ends of the band that enabled the wearer to tie it onto his or her head. Many of the grander ones were of gold and so lavishly ornamented that they qualify as jewelry.

There may be individual cameos with the portrait bedecked in jewels earlier than the Renaissance, but their numbers are few. From 1700 onward Italy excelled in carvings in which the female subject was adorned with jewels, often a string of beads around the neck or hair, sometimes beads and earrings or other combinations of jewelry. It is almost customary today for carvers to depict female figures with jewelry (fig. 4-15).

## DESIGN

One means that may help determine when a cameo was carved is an inspection of the accuracy—the integrity—of the design. If it looks too newly crafted to be an ancient or Renaissance piece, then it may not be that old. Every culture has had its own standards of what is considered accurate and beautiful. The definition of good design changes as the technology to execute it improves. This does not mean that every piece produced today with modern tools is a finely crafted piece with good design; it is also true that as the demand for goods rises and they become more common, a decline in the quality of workmanship inevitably results.

Cameo backs are frequently the subject of intense study and discussion. Some experts insist that an irregular but polished back is a mark of a Renaissance cameo. In those days the preparation of the rough was first carried out by an apprentice or workman, who smoothed the piece and finished the back before the stone was passed along to the master carver for design. This point has been disputed by carvers themselves, who argue that not all carvers had apprentices or workshop staff. While it is true that some antique cameos have polished and hollowed backs, some were carved especially for attachment to a garment or cuirass, and the back of the cameo is finished because these pieces were seen from the back as well as the front. Cameos with pierced backs usually were made especially to attach to material or other surfaces. Some, but by no means all, Hellenistic cameos are found with the backs hollowed out to conform to the contours of the front of the cameo, giving a kind of repoussé look to the carving.

## THE CUT

Few cameos can be dated to before the Augustan age. Those that do exist are very small, usually of low relief. The low relief resulted from the use of slow-cutting tools on hard materials. The greatest number of ancient cameos are small with flat surfaces. Their subjects often fill the field of carving, with very little background left between the figure or portrait and the edge of the cameo, because the objective of the carver was to eliminate as little as possible with his pointed tool. In *The Science of Gems, Jewels, Coins, and Medals,* Archibald Billing noted that it was not by will but necessity imposed by the tools, that the ancient workmen left almost no margins on their cameos.

The undercutting of cameo portraits or scenes to raise them into an almost three-dimensional relief has been done on and off throughout the history of the cameo. Various craftsmen would resort to this method for convenience instead of art because it saved them the time and trouble of leveling and polishing the backgrounds.

---

## SIGNATURES

Signatures are found on a number of old hardstone cameos, but their presence does not always determine the cameo's date because there is no guarantee that the signature is authentic. Many fine genuine antique hardstone cameos without signatures acquired them at a later time, from the hand of an entirely different carver. This practice persisted from the Renaissance into the nineteenth century; only when absolute, continuous provenance, supported by unquestionable documentation, is available can the signature on a cameo verify the work as that of a specific artist. The practice of adding a great master's signature to cameos that were not his was common because, as with any fine art, the signature significantly enhanced the value of a cameo. In commerce and art, signature forgery was a crime that did not go unpunished in sixteenth-century England, where the statutes dealing with the offense are legion. In 1562 in England, the punishment for forging a signature was at minimum a fine; at the most severe: pillory, having both ears cut off, the nostrils slit up and seared, forfeiture of land, and perpetual imprisonment. From 1634 the penalty was death.

In 1763 the German engraver Lorenz Natter wrote the *Treatise on the Antique Method of Engraving Gemstones,* listing all the examples he could gather on styles and signatures, but he confessed many more eluded him. In time Natter was himself accused of falsifying signatures, but he denied the accusation, claiming: "I am not ashamed to own that I continue to make copies (with Greek inscriptions and masters' names) at the times when I receive orders. But I defy the whole world to prove that I have ever sold one for antique."

Signatures prove to be an interesting exercise in deduction. On Etruscan scarabs, the name inscribed refers to the person represented there. On Greek gems the signature is the name of the artisan who made the item. In the second century B.C., the only concern of the Roman carver was to whom the piece belonged, and so inscriptions were of the name of the owner of the seal or cameo. The inscription could be in Latin, Greek, or Etruscan characters, but the majority named the owner, not its maker. Of cameos of the Greco-Roman period (Augustan and Early Imperial) that survived with an artist's signature, those made by the most distinguished artisans of the period were crafted by Dioskourides and his three sons, Eutyches, Hyllos, and Herophilos.

The following catalog of carvers is not to be considered comprehensive but rather contains the names of some of the most celebrated carvers who signed their work.

*Angelo Amastini* (1754 to 1815) and his son Nicolo were both known as fine gem carvers. Three cameos by Nicolo are in the Metropolitan Museum of Art in New York: the education of Bacchus, a free imitation of a Roman or Hellenistic composition on sardonyx, signed N. Amastiny in Greek letters; a maiden carved in classical dress, of two-layer onyx, signed Amastini; and Psyche with butterfly wings in her hair, on onyx, brownish white on a transparent brown ground, signed Mastini.

*Philipp Christoph Becker* (1674 to 1743) was born in Coblenz, Germany, and later worked in Vienna as a medalist and gem carver. Called at one time the best carver of gems in Germany, he did a portrait of Charles VI and cut many seals for German princes. From Vienna he was summoned to work in Saint Petersburg by Peter the Great. Much of his work involved coats of arms. Several gems of this period are signed "Hecker," probably the work of another artist, though C. W. King suggested the signature was a misspelling of Becker.

*Antonio Berini* (1770 to 1830) was born in Italy and studied gem engraving under Giovanni Pichler. Most of his subjects were classical, but at Windsor Castle there is an intaglio of Saint George signed by him. He is famous for one notorious cameo cut in 1802, a portrait of Napoléon which unfortunately had a red vein in the stone encircling Napoléon's throat. As a result of this foolhardy carving, Berini was imprisoned as a suspected assassin.

*William Brown* (1748 to 1825) was a British gem engraver active in Paris, London, and Saint Petersburg. There is a sardonyx cameo, white on brown ground, of Alexander Pope with a border inscription, Alexander Pope, signed Brown, in the Metropolitan Museum in New York.

*Edward Burch* (1730 to 1814) was a famous English engraver who also worked as a wax modeler.

*Alessandro Cades* (1734 to 1809) was an Italian from Rome who signed his full name, sometimes in Greek characters. Cades is mainly distinguished for his skillful copies of ancient cameos and classical carvers' signatures, commissioned by Prince Poniatowski to enlarge his collection.

*Giovanni Calandrelli* (1784 to 1852) was a Roman who became notorious for cameos' fakes. Many of his works were bought as antiques by the Berlin Museum.

*Giovanni Costanzi* (1674 to 1754) worked in Rome and Naples; two sons, Rommaso and Carlo, were also gem carvers. Giovanni worked on precious stones, especially diamonds, and is noted for having carved a diamond intaglio of Leda and the swan for the king of Portugal. His other famous clients were Empress Maria Theresa—she owned a sapphire cameo—and Pope Benedict XIV, who had an emerald cameo. Most of his works were of antique and classical subjects.

*Adolphe David* (1828 to 1896) was a sculptor, gem engraver, and medalist active in Paris. A cameo carved in three layers in sardonyx of Phaeton in his chariot pulled by four horses and signed A. David is in New York's Metropolitan Museum of Art.

*Johann Christoph Dorsch* (1676 to 1732) is said to have carved hundreds of cameo portraits of popes, emperors and kings, as well as many copies of antique gems. A daughter, Susan Dorsch, was also a gem carver.

*Julien de Fontenay* (unknown) carved the likeness of Queen Elizabeth I in cameos but signed very few of these. He is reported to have been the same carver known as "Coldore," a sobriquet given him at court, perhaps because of a gold chain that he wore ex officio, mentioned in the Lettres Patentes as the king's valet and engraver of precious stones. His cameos, all heads, were signed with the initials C. D. F.

*Francesco Ghinghi* (1689 to 1766) was a noted Italian carver, with numerous cameos in the Uffizi in Florence.

*Giuseppe Girometti* (1780 to 1851) was an Italian sculptor, medalist, and engraver of coins and gems, with numerous cameos on display in the British Museum. Three of note in the New York Metropolitan Museum are: Hercules with the lion's skin, a light red and brown sardonyx; an idealized portrait of a girl in pink on red-ground sardonyx; a bacchante bust in white on a grayish-red-ground onyx. All are signed Girometti.

*Jacques Guay* (1711 to 1743) was a Parisian who studied engraving and cameo carving in Italy. Many of his early copies of antique cameos were sold as genuine. He worked under the patronage of Louis XV and instructed Madame de Pompadour in carving cameos. Guay did a series of gem intaglios recording important events of the early part of Louis XV's dynasty (1715 to 1774); the series became a part of the royal cabinet of gems.

*Romain-Vincent Jouffroy* (1749 to 1826) studied in Naples and worked in Paris. He carved many antique-style cameos, three of which are in the Bibliothèque Nationale in Paris.

*F. Carol Lebrecht* (d. 1828), a German carver, went to Russia to head the mint in 1775. He gave instructions in carving to Empress Maria Fyodorovna, widow of Paul I of Russia. A well-liked and much-respected artist, Lebrecht served four Russian monarchs and was appointed councillor of state. A recipient of the Order of Sainte Ann, he founded a school of engraving in Saint Petersberg.

*Leone Leoni* (1509 to 1590), born in Arezzo, Italy, was a sculptor, medalist, goldsmith, and gem engraver active in Rome, Genoa, Milan, Parma, and Brussels. He was a noted wax modeler. A two-sided cameo in the Metropolitan Museum of Art is attributed to him, although it is unsigned. The cameo is white on red-ground sardonyx with Emperor Charles V and his son King Philip II on one side; the reverse shows Empress Isabella. A bronze cast of the cameo is in Munich in the Munzsammlung.

*Nathaniel Marchant* (1739 to 1816), born in England, went to Italy to study gem engraving and was appointed engraver to the royal mint in 1782. He was one of a few English practitioners of the glyptic arts, preeminent in his generation and considered the only rival of the celebrated Pichler. He is credited with over two hundred gems, cameos, and intaglios.

*Niccolo Morelli* (1779 to 1836), born in Rome, was active in the service of Napoléon. Two cameos signed Morelli are in the Metropolitan Museum. One is a pink, brown, and white onyx of a bacchante wearing a vine wreath; the other, a three-strata brown, white, and black onyx of a bacchante. The head of the bacchante with the garland of vine leaves was a favorite subject of carvers in the nineteenth century.

*Matteo del Nasaro* (unknown) of Verona carved cameos for Francis I of France. Nasaro so impressed the king with his carving skills that Francis bestowed a pension and court appointment upon him. The artist later became the chief engraver to the mint at Paris.

*Lorenz Natter* (1705 to 1763) of Nuremberg was particularly adept at imitating antiques. Some misuse of his abilities was suggested during forgery and fraud scandals, but he contended he never sold his work as antique. Natter sometimes signed his name in Greek characters and also often used a Greek word for serpent because, in German, Natter means serpent. It was not unusual for the Renaissance carvers to sign their names in Greek characters, a practice that continued late into the eighteenth century. The carvers of this era were greatly enamored of their predecessors' work and did all they could to equal them. A cameo by Natter of Lord Duncannon in antique style with the inscription W. Lord Duncannon, signed L. Natter F. is displayed in the Metropolitan Museum.

*Antonio Pazzaglia* (1736 to 1815) was a member of the papal Swiss Guard in the Vatican who became a gem carver; his forte was imitating antique gems, which he often signed with his name in Greek characters.

*Pier Maria Serboldi da Pesca* (1455 to 1522), also called I1 Taglicarne, was a goldsmith in Florence, artist at the papal court of Leo X, and engraver at the mint. He was celebrated for his talent in both carving and painting. He carved the signet *Festival of Bacchus* for his friend Michelangelo. The piece remains in the French cabinet.

*Anton Pichler* (1697 to 1779) and his sons, Giacomo, Giovanni, Giuseppe, and Luigi, spent many years as master engravers. Luigi was professor of medal art and stone engraving in Vienna. Many Pichler stones are signed in Greek characters (fig. 4-16); several are on display at the Metropolitan Museum. One, a portrait of a Roman couple by Giovanni, is the only known double portrait by him.

*Benedetto Pistrucci* (1784 to 1854), born in Rome, was one of the most celebrated cameo carvers of all time; his creative genius was often compared to that of Benvenuto Cellini. Pistrucci's autobiography, translated by Archibald Billing in *The Science of Gems, Jewels, Coins, and Medals,* is a fascinating account of the struggle to be recognized as an artisan in his time. Pistrucci enjoyed the patronage of kings and was appointed chief engraver at the royal mint in London. He designed an

FIGURE 4-16.

*A sardonyx cameo signed with the name Pichler. (Content Family Collection of Ancient Cameos, Houlton, ME)*

engraving of Saint George slaying the dragon for the British gold sovereign coin (fig. 4-17), which has recently been featured on a series of gold proofs. Pistrucci's engraving of Saint George slaying the dragon first appeared on the sovereign in 1817, and it has stood the test of time and scrutiny of generations, having been frequently used on British gold coinage, with only minor changes since its introduction.

But Pistrucci had critics, some of whom refused to believe that he was such a talented carver of antique-style cameos that they were virtually indistinguishable from the originals (fig. 4-18). There is the fascinating story of a small carving of the head of the goddess Flora. Pistrucci carved this cameo head with a wreath of roses for a very well known glyptic arts dealer of the time, Angelo Bonelli. The artist himself told of putting his mark under the setting near a broken edge of the stone. The crafty dealer Bonelli bought the Flora from Pistrucci for only a few pounds, but a later written account of the affair reveals he sold it for a huge profit to the collector Richard Payne-Knight. One day, while Pistrucci was working on a wax model of the head of Sir Joseph Banks in preparation for a carving, Payne-Knight burst into the Banks home, excited and anxious to tell about his extremely good fortune in buying the Flora cameo from Bonelli, who had sold it to him as a newly discovered antique. Pistrucci, startled to hear such a claim and anxious to set the story straight, pointed out to the men that he had cut the gem himself and showed them the secret mark, an inverted V, with which he branded all his works. Further, he asked them to

**FIGURE 4-17.**

*The Pistrucci engraving of Saint George slaying the dragon, which first appeared on the gold sovereign in 1817.*

FIGURE 4-18.

*An early-nineteenth-century onyx cameo 1¾ inches, carved in three strata, brown and white on a black ground, by Benedetto Pistrucci. The subject is Venus Marina standing in a shell drawn by two dolphins. In her left hand are the reins; at her right, a cupid. The cameo is signed Pistrucci. It is in a contemporary gold mounting. (Metropolitan Museum of Art, The Milton Weil Collection, Gift of Mrs. Ethel Weil Worgelt, 1940, 40.20.49)*

observe that he had carved a species of roses that was not known in ancient times. Shocked and agitated, Payne-Knight refused to believe the artist, even after a botanist confirmed that the particular variety of roses on the cameo was a comparatively new species. He persisted in his refusal to admit that he had been cheated to the end of his life. At his death he bequested the Flora, along with other cameos and engravings, to the British Museum.

Pistrucci's two daughters, Maria Elena and Maria Elisa, were also gem carvers. The Metropolitan Museum has cameos from all the Pistruccis. Benedetto Pistrucci signed his Pistrucci; Maria Elisa, M. E. Pistrucci; Maria Elena, E. Pistrucci.

*Philippo Rega* (1761 to 1833), a Neopolitan, worked under the patronage of Ferdinand I, the king of the Two Sicilies, as well as for court nobles. He was commissioned for cameo portraits by Sir William Hamilton and his wife, Emma, Lord John Hervey, earl of Bristol, and Admiral Horatio Nelson. A superior and highly detailed portrait of Zeus in gray and white onyx, signed Rega, is in the Metropolitan Museum. It is documented in a museum catalog as being "worked in a properly archaeological spirit, characteristic of the artist." Rega did portraits of Napoléon's brother Joseph and the rest of his family. He was also decorated by the French general Joachim Murat after Rega finished a cameo portrait of the hero.

*Giovanni Antonio dei Rossi* (1494 to 1540) of Milan was noted for a large cameo of a group portrait of Cosimo de Médici and his family. He taught Catherine de Médicis to carve gems.

*Giovanni Antonio Santarelli* (1758 to 1826) worked in Rome and was an active gem engraver and medalist in Florence and Milan as well. Early in his career he worked with the Pichlers. He became infamous as one of the carvers who did a multitude of forgeries for the collection of Prince Stanislaw Poniatowski. One of his cameo carvings, of two dancing cupids, on white and gray onyx, is in New York's Metropolitan Museum, signed Santarelli.

*Luigi Saulini* (unknown), a Roman native who worked in England, produced some of the most spectacular cameos of any era. His father, Tommaso (1793 to 1864), was also a renowned gem carver; they both exhibited at the International Exhibition in London in 1862. Luigi's portrait of Florence Nightingale in dress of the period was carved during the height of her fame, owing to her activity during the Crimean War as general superintendent of the English field hospitals. The cameo is in the Metropolitan Museum. It is a white-on-red sardonyx and is signed L. Saulini (behind the neck). Other cameos by Saulini in the Metropolitan Museum are an Apollo in white and black onyx, signed L. Saulini F.; the discobolos (fig. 4-19), a white and black onyx copy of the famous antique work by Myron, probably after the group in the Vatican, signed L. Saulini F.; Cupid and Psyche kissing, in white and black onyx, after the antique marble group in Florence, signed: L. Saulini; Cupid with the bow, in black on white onyx, signed, L. Saulini F.; a spectacular white on black onyx carving of the toilet of Nausikaa, signed L. Saulini (fig. 4-20).

*Louis Siries* (1686 to 1757) was goldsmith to Louis XV; in 1757 he published a catalog of 168 gems he had engraved. His specialties were cameos and intaglios scaled to microscopic proportions because he did not want to be accused of imitating antique works. He signed his name in Latin. Empress Maria Theresa also extended him her patronage.

*Flavio Sirletti* (1683 to 1737) was an Italian who excelled in cameo portraits and revived the practice of working with diamond points. He supposedly used diamond points to prove that such a carving method was possible. His work was delicate and beautiful, frequently taken for classical Greek work. He signed his name in Greek characters. It is known that he

FIGURE 4-19.

*The famous cameo necklace by Luigi Saulini called* The Discobolos, *in white and black onyx, signed L. Saulini. The 1 1/16-inch-high discobolos cameo to the left is a copy of the famous antique work by Myron, probably after the Vatican group. In the center is a cameo of Cupid and Psyche kissing, in black and white onyx, 1 1/8 inches high, after the antique marble group in Florence, signed L. Saulini. On the right is Cupid with a bow in black and white onyx, 1 1/16 inches high, signed L. Saulini. At the top is an onyx cameo, 13/16 inch high, of an athlete bearing a large hoop, after a famous classical intaglio. (The Metropolitan Museum of Art, The Milton Weil Collection, Gift of Mrs. Ethel Weil Worgelt, 1940, 40.22.55c)*

reworked many of the ancient gems and applied numerous bogus Greek inscriptions to original works of ancient times. A son, Francesco, was also a gem engraver.

*James Tassie* (1735 to 1799) invented a vitreous glass material that was suitable for imitating gems in 1763. The process was a jealously guarded secret, but it was known that the material was a fusible glass, which could be made to look like varied layers of chalcedony cameos. In 1769 Tassie supplied casts of gems to Wedgwood and in 1775 published a *Catalogue of Impressions in Sulphur of Antique and Modern Gems.* The glass, or paste, gems, were made from the sulphur casts and sold in great number. Another catalog of Tassie's collection, with 15,800 casts, was issued in 1791. The original works, all portraits, were taken from wax models, not from hardstones. His nephew William (1777 to 1860) added to the collection of reproductions and published three catalogs. That one published in 1830 listed 20,000 antique and modern gems.

*C1. A twentieth-century replica of the celebrated Tazza Farnese cameo by the master carver Richard H. Hahn of Germany.*

*C2. This Hellenistic four-layer sardonyx cameo, about ⅞ by ⅔ inch (22 by 17.5 millimeters), depicts an armed, helmeted warrior leaping from his horse with his shield which bears the head of Medusa. This cameo from the first century B.C. is set in a gold-enameled seventeenth-century mounting. Provenance: Ex Charles Newton-Robinson collection; ex Sir Robert Mond collection; published in the Burlington Fine Art Club "Exhibit of Ancient Greek Art," 1903, London. (Content Family Collection of Ancient Cameos, Houlton, ME)*

*C3. An A.D. 14 Augustan cameo in two layers of white on black onyx, mounted in an eighteenth-century gold ring. The subject is a silenus reclining on a lion skin draped over a cart, which is being pulled by two erotes. Psyche, with butterfly wings, is beside the silenus. The cameo is styled in the manner of Sostratos. It measures about ¾ by ½ by ⅛ inch (19 by 15 by 4 millimeters). Provenance: Sir Francis Cook collection, July 1925. (Content Family Collection of Ancient Cameos, Houlton, ME)*

C4. (left) *The famous Gemma Augustea, one of the masterpieces of ancient glyptic art, has been dated as first century A.D. The celebrated cameo has belonged to the church, to kings, and has been a trophy of war. (Courtesy of the Kunsthistorische Museum Vienna)*

C5. (left) *A late-nineteenth-century sardonyx cameo grylli with two faces, measuring about 1½ by 2 inches (40 by 54 millimeters). (Ann Spector Collection)*

C7. (right) *Blue and white high-relief chalcedony cameo of Jupiter (Zeus). He would have been identified as Saint John in the early Christian period. (From the collection of The Lizzadro Museum of Lapidary Art, Elmhurst, IL)*

C6. (above) *Every veiled woman was identified as the Madonna in the early Christian period, as in this early-twentieth-century blue and white hardstone cameo. (From the collection of The Lizzadro Museum of Lapidary Art, Elmhurst, IL)*

*C8.* (right) *Brown and white twentieth-century hardstone cameo of Leda and the swan. In earlier times its subject would have been interpreted as the Annunciation, to eliminate the pagan connotation. (From the collection of The Lizzadro Museum of Lapidary Art, Elmhurst, IL)*

*C9.* (left) *The revival of classicism during the Renaissance made Greek and Roman myths popular cameo subjects. (Ann Spector Collection)*

*C10.* (right) *Since earliest times reinterpretations of the Greek and Roman myths have engaged the skills of carvers. (Ann Spector Collection)*

*C11.* (left) *This display of early Greek and Roman myths carved in shell demonstrates the virtuosity of master carver Dante Sebastianelli, of Torre del Greco. (Ann Spector Collection)*

*C13.* (below) *The story of Eros, who was called Cupid by the Romans, and Psyche is above all a love story. The most famous telling of the story is in the* Metamorphoses *by Apuleius, written in the second century* A.D. *(From the collection of The Lizzadro Museum of Lapidary Art, Elmhurst, IL)*

*C12.* (above) *This depiction of the birth of Venus has elements taken from Nicolas Poussin's 1635 masterpiece. Numerous versions of the myth have been rendered, by many creative artists. (Ann Spector Collection)*

*C14.* (left) *Shell became an important medium for carving cameos because it was abundant and cheap. (Photo courtesy of Nino Tagliamonte)*

*C15.* (right) *These amusing late-twentieth-century shell cameo cupids mimic the antics of those depicted in the art of Pompeii during the first century* A.D. *(Courtesy of Di Vinci Ltd.)*

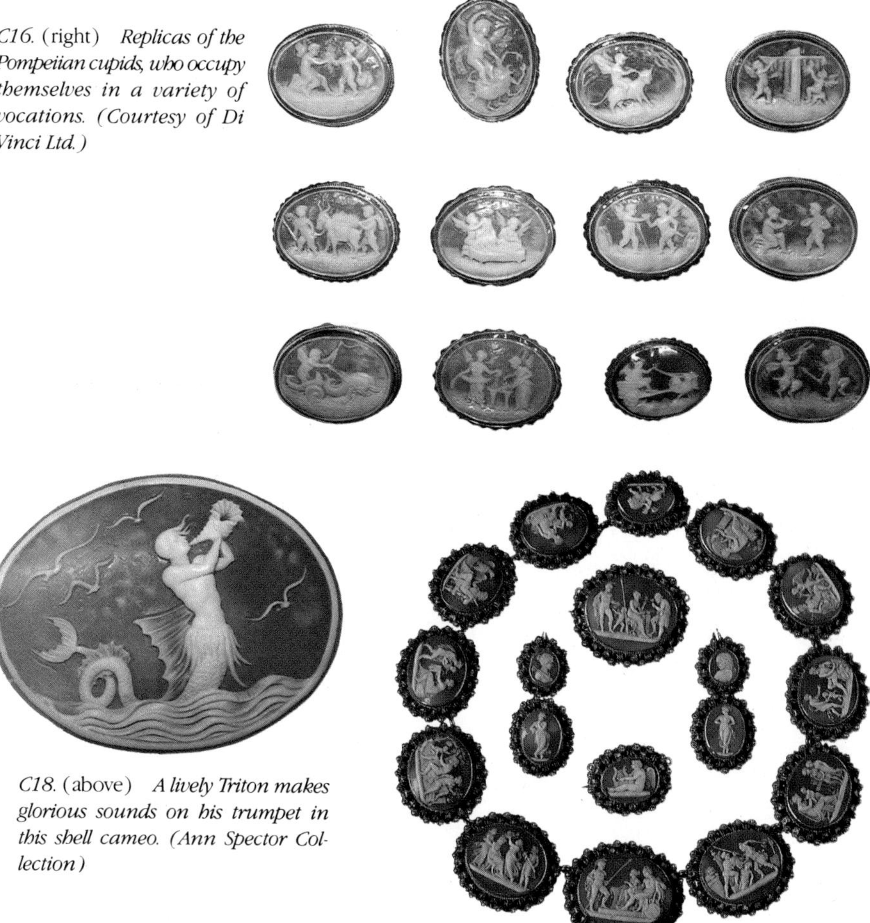

*C16.* (right) *Replicas of the Pompeiian cupids, who occupy themselves in a variety of vocations. (Courtesy of Di Vinci Ltd.)*

*C18.* (above) *A lively Triton makes glorious sounds on his trumpet in this shell cameo. (Ann Spector Collection)*

*C17.* (above) *A nineteenth-century carnelian shell cameo parure by the celebrated Italian carver M. Federici, nicknamed "The Prussian." (Photo courtesy of Basilio Liverino)*

*C19.* (left) *A carved cameo plate by Richard H. Hahn, 8 inches in diameter, portrays the three Fates in Brazilian agate. Clotho with her spindle spins the thread of a man's life; Lachesis measures the length of life; and Atropos, with shears, cuts the thread of life. (From the collection of The Lizzadro Museum of Lapidary Art, Elmhurst, IL)*

*C20.* (right) *A Medusa head carved in Torre del Greco from bright red coral and made into a brooch accented with diamonds. (Ann Spector Collection)*

*C21.* (left) *A three-layered agate bowl picturing the judgment of Paris. (From the collection of The Lizzadro Museum of Lapidary Art, Elmhurst, IL)*

*C22.* (below) *A shell cameo carved after Titian's* Bacchanal, *circa 1518. Modern carvers often take elements from separate artworks, combining them to create new variations on a subject. In the Bacchanal the god of wine, Bacchus, created a stream in which wine flowed and the people lived in a state of happy intoxication. In this motion-filled cameo, the carver has portrayed a nude who sleeps, satiated with drink; a male in the center pours wine, and dancers echo movement. (Courtesy of Gennaro Borriello s.r.l.)*

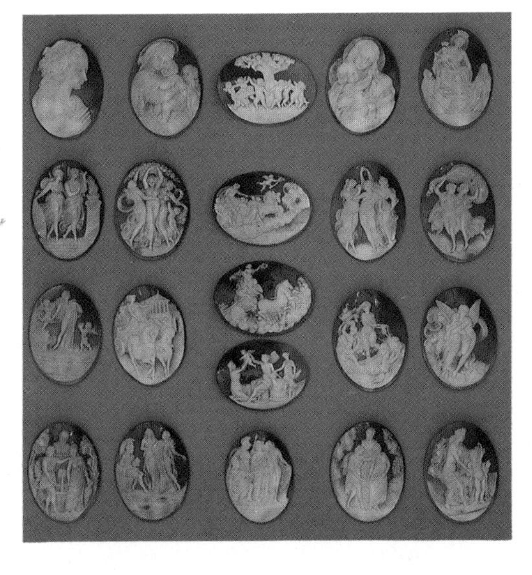

*C23.* (above) *Shell carvers today reinterpret, select, and create subjects. (Photo courtesy of the Italian Institute for Foreign Trade)*

*C24. An abstract cameo from Richard H. Hahn. (Copyright F. A. Becker)*

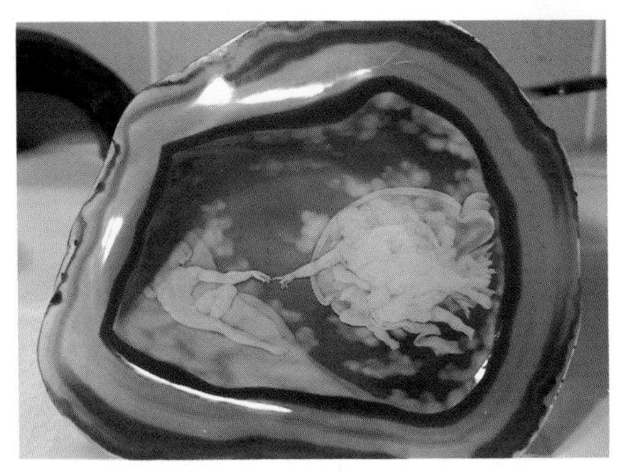

*C25. The ceiling of the Sistine Chapel was the inspiration for this agate cameo dish, in the Deutsches Edelsteinmuseum, Idar-Oberstein.*

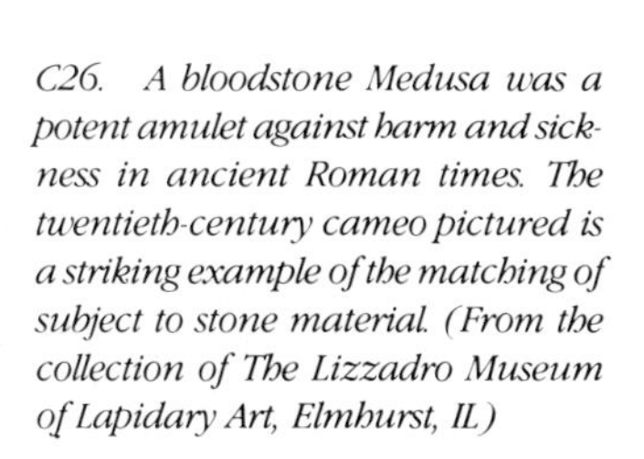

*C26. A bloodstone Medusa was a potent amulet against harm and sickness in ancient Roman times. The twentieth-century cameo pictured is a striking example of the matching of subject to stone material. (From the collection of The Lizzadro Museum of Lapidary Art, Elmhurst, IL)*

*C27. The hand-carving of hardstone cameos requires creative as well as technical skills. (Copyright F. A. Becker)*

*C28. Master carver Richard H. Hahn at his workbench. (Copyright F. A. Becker)*

*C29. In hardstone carving the stone is moved against the electric cutting tool. (Copyright F. A. Becker)*

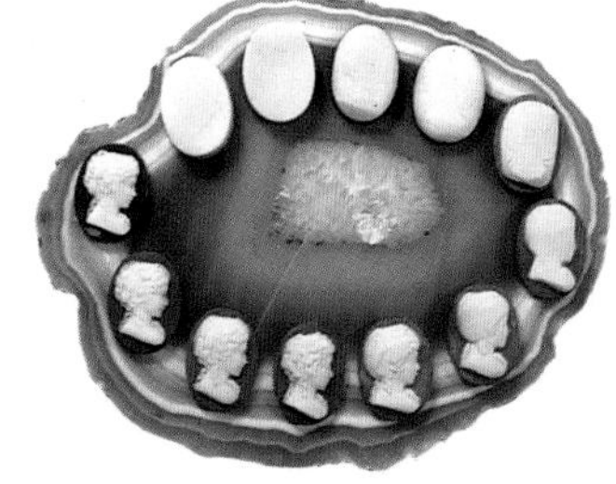

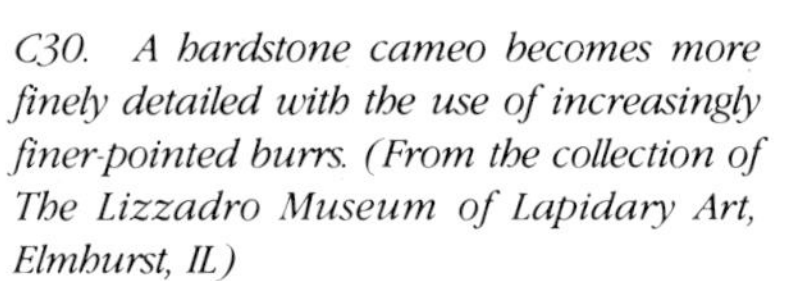

*C30. A hardstone cameo becomes more finely detailed with the use of increasingly finer-pointed burrs. (From the collection of The Lizzadro Museum of Lapidary Art, Elmhurst, IL)*

*C31. The only electric tool used by a shell cameo carver is a dental drill.*

*C32. Aray Dommer, an Idar-Oberstein master carver, prepares a cameo from a photograph.*

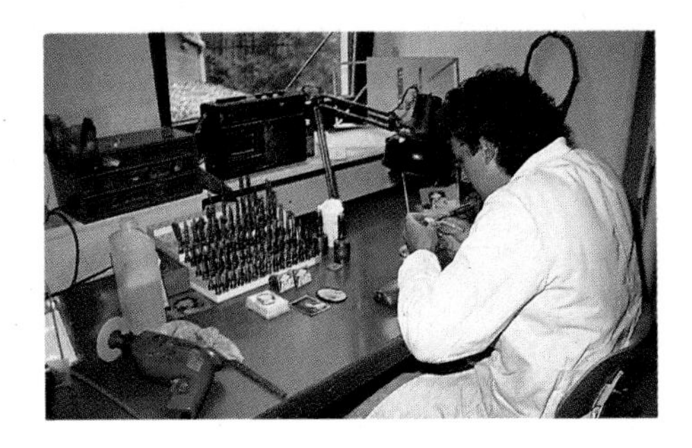

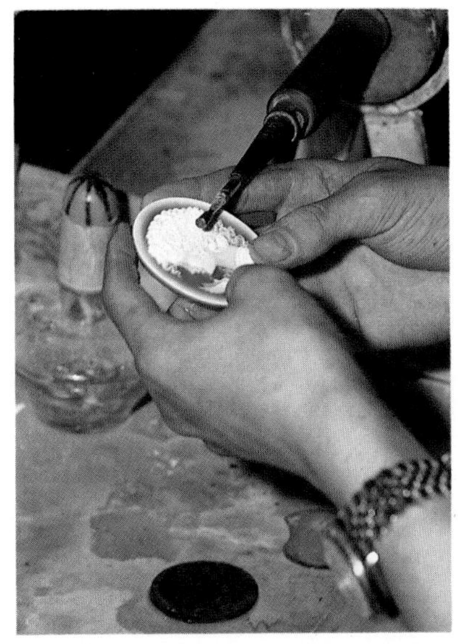

*C33. Ultrasonically carved cameos are often hand-finished for greater salability. (Photo courtesy of Pointers Gemcraft Manufactory)*

C34. (left) *An example of undercutting, where the subject is in high relief and cut, almost in the round, away from the background. (From the collection of The Lizzadro Museum of Lapidary Art, Elmhurst, IL)*

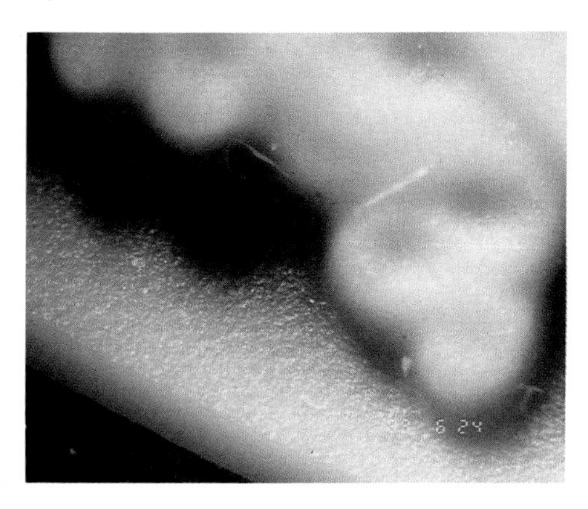

C36. *The FFS syndrome, an absence of any undercutting or tool marks, a matte finish, and a repetitive design are all clues to identifying the ultrasonically carved cameo. This magnified look at the white carved layer reveals the fresh-fallen snow look.*

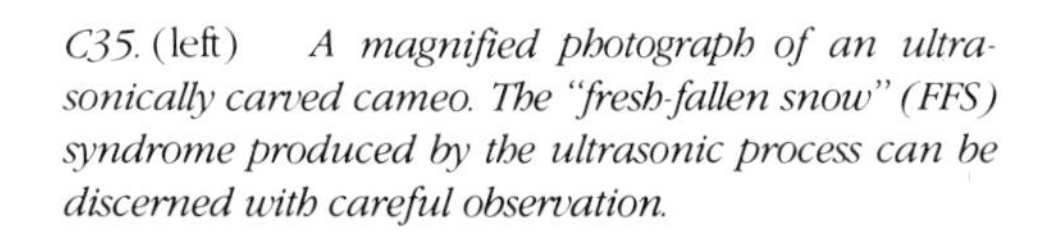

C35. (left) *A magnified photograph of an ultrasonically carved cameo. The "fresh-fallen snow" (FFS) syndrome produced by the ultrasonic process can be discerned with careful observation.*

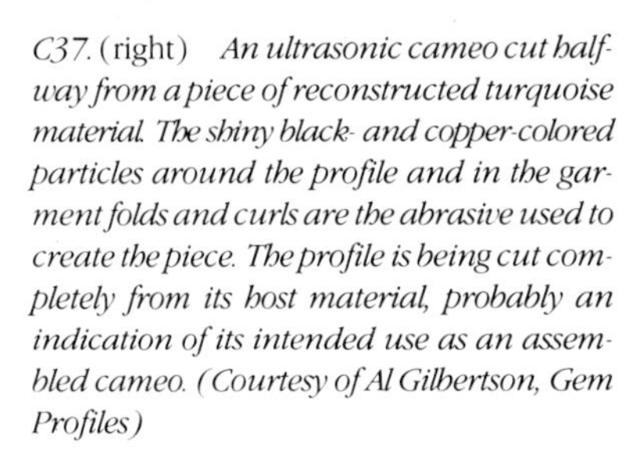

C37. (right) *An ultrasonic cameo cut halfway from a piece of reconstructed turquoise material. The shiny black- and copper-colored particles around the profile and in the garment folds and curls are the abrasive used to create the piece. The profile is being cut completely from its host material, probably an indication of its intended use as an assembled cameo. (Courtesy of Al Gilbertson, Gem Profiles)*

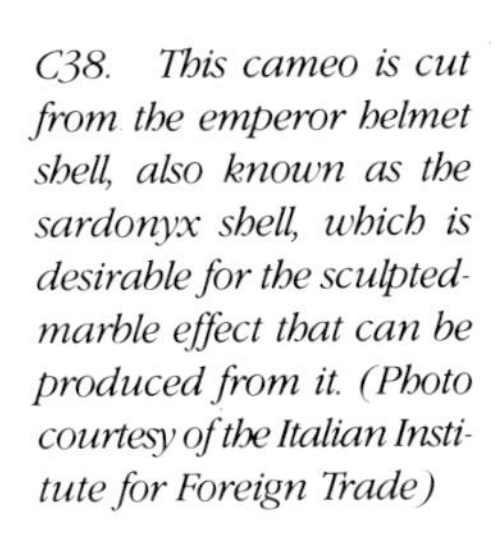

*C38. This cameo is cut from the emperor helmet shell, also known as the sardonyx shell, which is desirable for the sculpted-marble effect that can be produced from it. (Photo courtesy of the Italian Institute for Foreign Trade)*

*C39. Some shells are complete works of cameo art themselves. (Photo courtesy of Nino Tagliamonte)*

*C40. When the carver begins work on the shell blank, it is attached to a dopstick for easier handling.*

*C41. The classic female profile on a shell cameo, created by the hands of a Torre del Greco carver. (Photo courtesy of the Italian Institute for Foreign Trade)*

*C42. Using as much of the white layer as possible, the carver transforms the shell cameo into a miniature sculpture. (Cameo courtesy of the Sebastianelli company)*

*C43. Cameos in shell that depict ancient legends and myths. (Courtesy of Gennaro Borriello s.r.l.)*

*C44. The city of Idar-Oberstein, Germany, stone cameo-carving capital of the world.*

*C45. Grinding wheels were operated by workmen lying on their stomachs and pressing the stone against the wheel. (Copyright F. A. Becker)*

*C46. Torre del Greco, at the foot of Mount Vesuvius, on the beautiful Bay of Naples. (Photo courtesy of Nino Tagliamonte)*

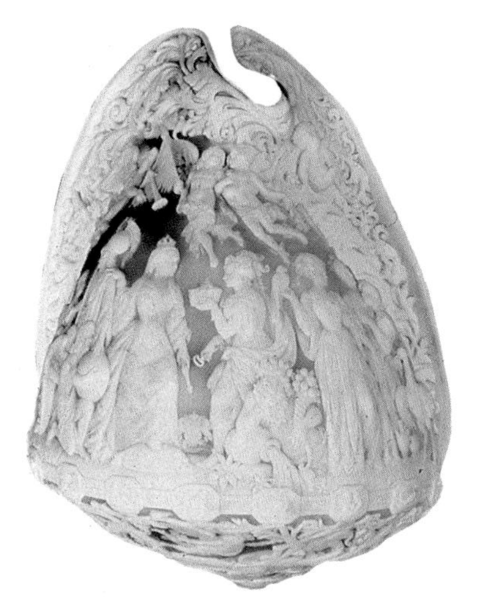

*C47.* The British Empire, *with Queen Victoria and allegorical figures of the empire in sardonyx shell. (Photo courtesy of Basilio Liverino)*

*C48. At the turn of the twentieth century, during the art nouveau period (1895 to 1915), sensuously carved female portraits were in vogue. (Courtesy Dr. Joseph Sataloff)*

*C49. Athletes and theater personalities were popular cameo subjects in the early 1900s. This shell cameo portrays Helen Wills Moody, tennis champion. (From the collection of The Lizzadro Museum of Lapidary Art, Elmhurst, IL)*

*C50. Victorian hairstyles are depicted as long and loose, sometimes with ringlets or long curls, usually adorned with flowers or jewels, bonnets or ribbons. This Victorian-style ivory cameo measures about 1½ by 2 inches (40 by 54 millimeters). (Ann Spector Collection)*

*C51. An ivory and malachite assembled cameo. (From the collection of The Lizzadro Museum of Lapidary Art, Elmhurst, IL)*

*C52. A contemporary carved cameo cup that is agate, not glass, produced in Idar-Oberstein and presently on display in the Deutsches Edelsteinmuseum. Cameo glass is carved in the same way as natural-layered stone.*

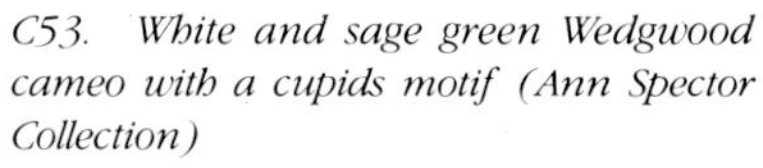

*C53. White and sage green Wedgwood cameo with a cupids motif (Ann Spector Collection)*

*C54. A cameo wax portrait of the duke of Wellington. (From the collection of Mrs. Muir Rogers)*

*C55. A well-preserved cameo wax portrait of Empress Catherine II of Russia. (From the collection of Mrs. Muir Rogers)*

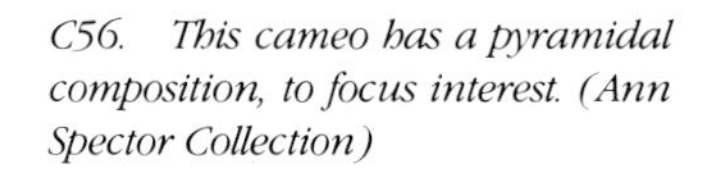

*C56. This cameo has a pyramidal composition, to focus interest. (Ann Spector Collection)*

*C57. This large hardstone cameo, about 2 by 2½ inches (49 by 67 millimeters) has balance and a well-contrived arrangement of figures, giving equilibrium to the scene. (Ann Spector Collection)*

*C58. This shell cameo by the Sebastianelli firm displays a sense of movement.*

*C59. The thin white layer on sardonyx, preferred by many carvers today, is often a fraction of the depth of a small match. (Copyright F. A. Becker)*

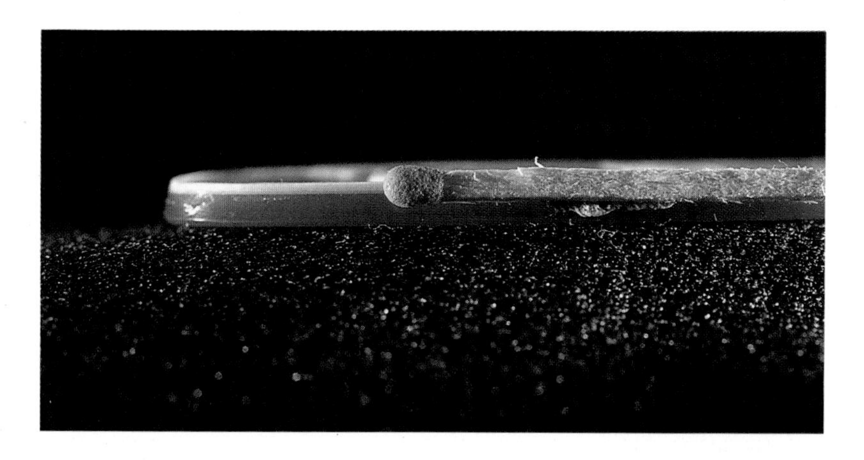

*C60. A remarkable shell cameo from the J. Daniel Willems collection. Called the "countess" by Willems, it is after a painting in the Museo Poldi Pezzoli in Milan. (From the collection of The Lizzadro Museum of Lapidary Art, Elmhurst, IL)*

FIGURE 4-20.

*An extraordinarily carved black and white onyx cameo by Luigi Saulini from the early nineteenth century, entitled* Toilet of Nausikaa, *in a gold tiara setting by Castellani. In the cameo Nausikaa is seated facing left; before her are a girl offering her a necklace and another kneeling with a mirror; behind Nausikaa are a girl dressing her hair and another arranging a basket; at the end is an altar. (The Metropolitan Museum of Art, The Milton Weil Collection, Gift of Mrs. Ethel Weil Worgelt, 1940, 40.20.55a)*

And finally, this brief review of engravers must also include the name of *D. Calabresi,* who is but a footnote in any book on cameo history but must have been an incredible craftsman. Or, at least, Pope Gregory XIII thought so in the sixteenth century. Calabresi was condemned to life imprisonment for murder, but so great was his fame as a gem carver, so delightful and skillful his work, that Pope Gregory XIII remitted his sentence when Calabresi consented to engrave for him a gem with a classical subject. The power of art!

## DETECTING CAMEO FAKES, FRAUDS, AND FORGERIES

The flip side of gaining expertise in dating cameos is having the ability to recognize fakes and forgeries.

Two primary types of cameos were produced in the early eighteenth century: fraudulent copies of antique models and frankly modern carvings of allegorical and historical scenes, and portraits. Although some copies, even forgeries, of earlier cameos and seals existed before the Renaissance, collecting of the antique gems and cameos, at first a fashionable amusement, which grew into an uncontrollable mania in the eighteenth century, created a fertile ground for the dishonest. Because of the cameo's popularity, legitimate dealers were demanding enormous sums of money for both the well executed and the crudely carved. When the art forger saw the seemingly insatiable public appetite, the fakes went into full production.

Most forgeries reveal themselves in time, but it may take the next generation to see the distinguishing stylistic traces left by the artisan. When the forger is poorly educated about the period he is copying, the result will be inaccurate inscriptions, lettering styles, mannerisms, or codes of dress of the time and place he has chosen to replicate.

The ancient classical cameos from Greece and Rome are usually of sardonyx but rarely in a high bas-relief. The high bas-relief was a practice of the Renaissance lapidaries; this same characteristic is found in the eighteenth-century forgeries of the original work of the fifteenth century.

Scientists have almost no means of detecting forgeries, and the deduction must remain an undertaking by the aesthetically sensitive collector or curator. Some scholars think they can detect forgeries by examining the strength of the carving, reasoning the lines and form executed by the hands of the master carver are strong, while the forger repeats the lines with hesitancy, as sketches. This method is totally unreliable, however, because cutting techniques have changed so slightly over the centuries. It is certain, though, that original antique pieces replicated in the eighteenth and nineteenth centuries often show visible file marks that appear too sharp to be taken for old work. Also, the uniformity of detail may provide some information about forgeries. An ancient worker did not use a great amount of detail in one part of his work, for instance, the hair or garment, but not in the other parts. Any variation in details, a lavishing of attention on one feature while leaving the remainder unfinished, is typical of the nineteenth- and twentieth-century carver.

Some slight clue may be found in the polishing finish, because it is known that most ancient stones were never excessively polished. However, one cannot rely on a well-worn appearance as evidence of the archaic.

There are many accounts of acid being used on stones, seals, and cameos to give them an appearance of age. Submerging onyx in an acid or lye bath will give it a dull, glossless look, and numerous writers have repeated the highly creative method of aging stones by using the stone-in-the-chicken ploy. This scheme calls for stuffing the newly carved or engraved stone down the gullet of a turkey or chicken so that the abrasive action of the gizzard imparts a dull and chalky "antique" look to the stone. Another method calls for shaking the stones in containers of sand to create a scratched surface, giving an aged look. Many gems have been purposely broken into fragments to make them look archaic and time-worn.

A favorite trick of the dishonest carver or seller is to surround the cameo with what seems to be solid provenance. The cameo is offered for sale with a bogus list of former famous owners or some allusion to a historical event. The detection of fraudulent schemes in this market amounts to a battle of wits, requiring a keen understanding of the art form and knowledge of art and human nature on the part of the collector or scholar. While it is helpful to understand the iconography of the past, if the carving is a direct copy of an original work of the period, it may not have any iconographic faults. If the buyer feels instinctively that the cameo's subject is out of sync with the culture of the proposed period, he also must know why. It may be an impossible costume, a strange hairstyle or headdress, a stilted pose, a wrong attribute, a faulty inscription, an incorrect spelling, or a multitude of other small inconsistencies.

Being able to detect fakes does not end with the individual collector. Museum collections are equally at risk. It is with the exposure of forger's tricks in mind that some museums have exhibited fakes alongside genuine articles. The Metropolitan Museum of Art in New York put on an educational display about which Gisela Richter, then curator of the classical department said, "The fakes are getting more clever all the time. . . . It takes a real artist in his own right to make a good forgery. The good forger is an artist who is also an archaeologist."

The University of Pennsylvania Museum has an extraordinary array of cameos comprising the majority of the Maxwell Sommerville (1829 to 1904) collection. Sommerville was professor of glyptology at that university and bequeathed to the museum one of the most important collections of ancient and neoclassical cameos and intaglio-engraved gems in the United States. The controversial nature of the collection has simply stressed its importance in the glyptic arts field. On one hand the collection has been widely acclaimed by scholars as the rival of cameo collections in Naples, Florence, Paris, Berlin, Vienna, and Leningrad. But it has also been called spurious by the most famous of glyptic experts, the eminent German archaeologist Adolf Furtwängler. In the fall of 1904, Furtwängler arrived from Munich to spend a few hours at the museum, briskly touring the collection. He raised serious doubts about some gems' authenticity, pronouncing a number of the cameos fakes or forgeries in a terse list prepared after his visit. Even though he did not have the opportunity to inspect all the gems in the collection, his criticisms were blunt and stinging: "All forgeries"; "Genuine but inferior specimens"; "Are nearly all genuine but quite common"; "Ordinary antique"; "Antique but late and bad"; "All in the case spurious." To the credit of the collection, Furtwängler did list numerous as genuine, but he gave no plaudits. He reported one ancient Etruscan specimen was a "very fine and rare specimen," and one was praised as "excellent," but no other measure of approval was extended.

One of the most treasured of the Sommerville cameos is a dark reddish-amber chalcedony, with slight white, sepia, and dark brown spots, or maculation, about 4¼ by 5¾ inches (11 by 15 centimeters), called *The Triumph of Constantine* (fig. 4-21). Professor Furtwängler is reported to have denounced it as "a very bad forgery."

Sommerville claimed the cameo was engraved in Byzantium during Constantine's rule of the Roman Empire. He said it had fallen into the possession of the Russian empress Catherine the Great, who gave it as a special reward to one of her ambassadors in 1785. The cameo was later traced to Greece, where it had been jealously guarded by its owner and his heirs. Sommerville reported that he came into possession of the prize only after more than six years of negotiations with its owners. Sommerville tried to get positive authentication for the piece and carried on a steady flow of correspondence with C. W. King, professor at Trinity College and a well-

FIGURE 4-21.

*The celebrated sardonyx cameo known as* The Triumph of Constantine, *dated by Professor Maxwell Sommerville as Roman, from the fourth century* AD. *This cameo bears a striking resemblance to one pictured in C. W. King's* Handbook of Engraved Gems *that was in a collection in Vienna in the late nineteenth century. (The University Museum, University of Pennsylvania [neg. #56154])*

known glyptics expert. King wrote to Sommerville, saying of the cameo, "It is by far the most important of all similar works of the Lower Empire hitherto published" — high praise from the expert and definitely in opposition to Furtwangler's later cry of forgery. The authenticity of the cameo has never been satisfactorily resolved, but it was put on public display in an IBM Gallery of Science and Art exhibition in New York in 1989, where it was listed in the brochure as a fourth century A.D. Roman sardonyx.

Sommerville described the subject matter of the contested cameo in the following manner. The subject represents Emperor Constantine being crowned by Victory, standing behind him as he rides in a chariot pulled by four horses that are led by a soldier. Behind the chariot is Crispus, son of Constantine, clad in a toga, and Fausta, wife of Crispus, in full dress. Both figures are carved in full frontal style. At the other end of the scene is Helena, mother of Constantine, and beside her the soldier guiding the horses. Sommerville pointed out that there are several "evidences" of the genuine antiquity of the piece, for instance, "if we consider the horses of the quadriga, (they all) exhibit the true Roman stiffness of movement." He also pointed out that the historical figures have the low relief that invariably stamps the works of the period, regardless of the size of the cameo; also, the costume of the soldier leading the horses is without the conventional Homeric breastplate; instead he wears a thick tunic of apparently quilted material and a helmet on his head. This is a particular characteristic of the later period of the Roman system, when all armor was discarded by the infantry and one's defense was reduced to a shield. Sommerville observed that in works of the Médici period, for instance, an artist gave antiquity to a subject by carving armor on the soldiers. Sommerville paid special attention to the dress of the royal family and felt it conveyed the idea that the artist was contemporaneous and had copied the garments from life. He puzzled over the roll of paper (called a volumen) held in the emperor's raised right hand and speculated that the roll was the Book of Gospels, which indicated the source to which the victor gives the triumph that the gem perpetuates. In his final analysis of the cameo, Sommerville remained fully convinced of its date, origin, and his interpretation of design and wrote: "In point of historical interest the gem — Triumph of Constantine — ranks next to the Gemma Augustea."

Dozens of glyptic arts writers have remarked upon the stiffness shown in subjects in genuine Roman carvings, as Sommerville pointed out in his gem. Therefore, if stiffness is a clue to authenticity and dating, it is logical to assume that if subjects in purportedly archaic works convey action and drama, the work should be immediately suspect as a forgery.

One practical reason the forger may have made the shift from a statuelike pose to a suggestion of movement is because the public preferred it, and therefore, it was more salable.

A villainous character of the eighteenth-century forgery game was Baron Philipp von Stosch (1691 to 1757), a Hanoverian agent paid by England to spy on the movements of Prince James Edward, the Stuart Pretender, in Italy. Stosch had an insatiable desire for engraved gems and cameos and owned a cabinet filled with 3,544 carved gems, some genuine antiques and many forgeries. The huge accumulation was of such great interest to connoisseurs of the glyptic arts that in 1760, three years after his death, the German archaeologist Johann Winckelmann wrote a catalog describing the collection for Frederick the Great, the king of Prussia. Later the Stosch gems were acquired by the Berlin Museum, where they remain today.

Stosch built his collection by commissioning Natter and other celebrated artisans of the time to engrave gems in the antique style and add real or fictitious Greek names to their work or to previously unsigned genuine antiques. He then invented colorful historical provenance for them; each story of the origin of a piece is a jewel in its own way. The stories were so convincing that Stosch sold hundreds of cameos to royalty and nobility, as well as collectors throughout Europe. He kept certificates of sale along with their manufactured provenances.

A fascinating story about the lengths to which the crafty baron would go to obtain a piece for his collection is told by the curator of the Bibliothèque Royale in Paris. It revolves around a cameo of a bacchanalian festival engraved by Pier Maria da Pesca, one of the leading carvers in the Renaissance, who worked in Rome for the Pope Leo X. The stone was a masterpiece of miniature carving, with eleven figures in bas-relief in a festival scene. The carved gem was given by Pesca to his friend Michelangelo and was used by the latter as his signet.

Later the cameo fell into the hands of the King Louis XIV, who put it into the Bibliothèque Royale in Paris for safekeeping. It narrowly escaped being stolen by the wily Baron Stosch. As the story goes, one day the curator of the bibliothèque was showing the baron the royal carved gems when he observed that suddenly the famous gem was missing, just after the baron had inspected it. The curator immediately sent for a strong emetic, and when it arrived he forced the German to swallow it. The irritant worked quickly, and during a spell of vomiting the baron gave up the gem into a basin that was being held in front of him. History makes no mention of what happened next, whether Stosch was arrested for attempted theft or merely escorted from the hall.

## Discovering the Fakes with Forensics

There are few individuals using sophisticated laboratory equipment such as that at the Rathgen Laboratory in Berlin to expose fakes and forgeries. The Rathgen uses atomic absorption spectrophotometry on dissolved metal, analyzing the makeup of the alloy to determine if the mixture is appropriate to the object's purported age. Of course, this test is for metals, not stones. The scanning electron miscroscopy recently done in other laboratories, however, shows some differences between the abrasions on ancient cylinder seals and their modern, faked counterparts. Tell-tale signs on fakes are: edges are perfect, with no indication of age or wear; and patina, naturally attributed to burial or great age (on glass and some stones), is also missing.

Josef Riederer, professor of archaeometry at the Rathgen, uses technology to determine age, materials, and production properties of all types of art for the express purpose of exposing fakes and forgeries. An interesting revelation is that new marbles used in sculptures and bas-reliefs fluoresce as red-violet under ultraviolet light; old marble in the ancient sculptures usually has a spotty bluish-yellow color. Other bas-relief and statuary in bronze that are touted as Egyptian, Greek, or Etruscan are found to be certain frauds if they show a high concentration of zinc under analysis; zinc was not a component used in copper alloys until Roman times.

According to Thomas Hoving, the former head of New York's Metropolitan Museum of Art, forgeries occur in cycles of ten years. In an article several years ago, when questioned about common art forgeries and asked if a fake can remain undetected forever, he answered that almost every art fake is eventually discovered; that is probably also true of glyptic art. The sad consequence of fraud is that when the deceptions are revealed, the public's appetite for collecting often ends. Hoving explains this as a normal part of the collecting cycle: "I can only say that collecting is a love affair, and when you find out your lover has been deliberately lying to you for years, it's all over."

That was the precise scenerio in 1839 when, during the height of the cameo-collecting mania, a scandal of such proportions was revealed that it resulted in the sudden and nearly total collapse of the market.

## The Poniatowski Affair

No more fascinating story in the history of cameos exists than the fraud perpetrated by Prince Stanislaw Poniatowski, who inherited a remarkable, rare, and important collection of 154 antique engraved and carved gems from his uncle Stanislaw August Poniatowski, the last king of Poland. The king had carefully assembled his collection; every one of the 154 pieces had been acclaimed as a masterpiece of workmanship, and all were unquestioned as to their antiquity. One of the pieces was a peerless intaglio by the master Dioskourides, with provenance assigning it to one of the first great gem collectors, King Mithradates of Pontus in the first century B.C.

To his inheritance of 154 outstanding gems, the prince added nearly 3,000 more in antique style that he commissioned by contemporary Roman carvers Giuseppe Girometti, Nicolo Cerbara, and Tomaso Cades, who spent years in his service. The prince, well educated in the classics, enjoyed suggesting subjects for carving, mostly drawing from classical history or mythology. The carvings were apparently made with the specific intent to deceive, because at Poniatowski's direction, the Oriental sardonyxes, amethysts, and crystals were all signed, not with the names of the Romans who carved them, but with the names of real or imaginary ancient artisans.

The prince died at Florence in 1833, and his collection was offered at auction by Christie's in London on April 29, 1839, in an event that drew interested collectors from all over the world. The sale, covering 2,639 gems, lasted for seventeen days, and as the gems began to be scattered all over Europe, experts and connoisseurs started to complain of forgeries. Scholars and historians were embarrassed that they had so easily been hoodwinked by the contemporary Roman craftsmen and Poniatowski. Reviewing the collection before the prince's death R. Rochette wrote: "The collection is full of works by Pyrgoteles, Polyclites, Apollonides, Dioskourides, in greater numbers than there were in antiquity itself." And the Berlin curator of gems E. Tölken seemed to express more amusement than shock at the similarities in engraving styles by the so-called Greek and Roman engravers: "Pyrogoteles works like Evodus, and there are more than four hundred years between them."

Although most of the works had been done with considerable technical skill, one of the sad effects of the scandal was that genuine cameo antiques declined in both monetary value and prestige as collectible objects. In the mid-nineteenth century, about a hundred pieces from the collection found their way to the United States and were presented to the New York Charity Organization Society, to be sold for its benefit. Tiffany and Company, acting as agent, sold them for a fraction of their former value.

In retrospect, today's jewelry historians seem surprised that the affair was not brought to light earlier, claiming that there is no suggestion of antiquity in the dramatic flamboyancy of the Poniatowski creations. Although C. W. King wrote that the gems were easily spotted by "display of too much of the flighty Louis XV manner, even in the attitudes of the persons and the treatment of the drapery," he also praised the portraits and single figures as being closely aligned with the "antique spirit" of earlier works. Judgments of character and authenticity are made more easy, however, if the critic is separated from the object by time and culture. In the early nineteenth century, identification of a gem's origin was not always routine, and dissemination of information was not as efficient as it is today.

That the forgeries themselves were excellent works of art in their own right cannot be doubted. If the workmen who carved the Poniatowski fakes had signed their own names instead of forging the signatures of others, the pieces would be honored today as being among the finest of nineteenth-century works. As monuments in the history of art forgery, they have a special value and already are collector's items.

A few good craftsmen—notably the two daughters of Benedetto Pistrucci in Rome and Luigi Isler in Paris—struggled on after the forgery debacle was over, but by then the passion for collecting antique engraved gems by connoisseurs and the public had virtually disappeared. It has never revived with the vigor of that time.

An interesting account of the marketing of fake classical cameos in the eighteenth and nineteenth centuries is given by Judy Rudoe in the British Museum catalog for the exhibition *Fake? The Art of Deception.* Rudoe quotes a letter from the English sculptor Joseph Nollekens about Thomas Jenkins, a British gem dealer living in Rome in the 1770s:

*As for Jenkins, he followed the trade of supplying the foreign visitors with intaglios and cameos made by his own people, that he kept in a part of the ruins of the Coliseum (sic), fitted up for 'em to work in slyly by themselves. I saw 'em at work though, and Jenkins gave a whole handful of 'em to me to say nothing about the matter to anybody else but myself. Bless your heart! He sold 'em as fast as they made 'em.*

Certainly, it must be remembered that false gems are not a modern invention. There is a story about the Emperor Gallienus, who reigned in Rome from A.D 253 to 268. When a jeweler sold his wife, Empress Salonina, fake gemstones, she was more than a little distressed and asked the emperor to do something about it. The emperor immediately ordered the jeweler to be thrown to the wild beasts in the Roman arena. As the jeweler stood trembling and naked, awaiting his doom in the dirt of the enclosure, the door of the cage to the wild beasts was thrown open—and out came a rooster! The ironic comment of the emperor was: "He who cheats others should be cheated himself."

A much harsher punishment was meted out in India in the sixteenth century, where it was decided, "The vile man who fabricates false diamonds will sink into an awful hell, charged with a crime equal to murder."

Along with outright forgeries and the modification of authentic pieces to make them seem more valuable, for instance, by the addition of a signature, some cameos fall into the category of unintended fakes, reproductions that have changed hands several times and become confused with antique pieces.

Professionals guard against fakes by purchasing with great care. They judge the source of the antique, look at the history, and use touch and sight. They feel for undercutting, repairs, and changes in the texture of the surface. Some common-sense rules for the cameo collector to use in trying to avoid the fake antiques include the following:

- Every piece should be assumed to be new, faked, or replicated.
- If a piece looks brand new, it probably is.
- Marks or lettering on any carved gem should have a logical explanation or meaning.
- Magnification should be used to inspect surface scratches. A perfect surface may indicate the carving is new or recently repolished; uniform scratches over a polished surface may be an indication of artificial aging.
- Broken edges covered with a mounting give the appearance of careful preservation.
- Ancient motifs should be checked for proper use and interpretation, given the piece's alleged time and place of origin
- An inscription out of context with the language or culture of the period it purports to represent is a warning sign of fakes.
- A mixture of Greek and Roman (Latin) letters is an obvious sign of forgery.
- Symbols and attributes must be scrutinized to be sure they are not out of context and are presented normally. For instance, musical instruments should be technically correct: ancient carvers would have known that the lyre was always held in the left hand and the right hand either remained empty or held an object used to pluck the strings of the lyre.

If purchase of an expensive antique is contemplated, buy only from a well-known and reputable dealer who can be held accountable for the merchandise.

One must not presume that the ability to detect forgeries can be mastered by reading a book on the subject. All collectors and many museum curators, even those with years of expertise and wide-ranging connoisseurship, have at one time or another been taken in by a forgery of some type of art.

The completely infallible test for authenticating antique hardstone or shell cameos has yet to be invented. At present no modern technology, machine, or device can eliminate the need for the judgment of the knowledgeable collector.

---

## SHELL CAMEO FAKES

Today an imitation shell cameo that is mass-produced by plastic injection molding is widely marketed. For those individuals not well acquainted with distinguishing glass or plastic from shell, the cameo may look strikingly genuine upon first inspection. To detect this fooler, look at the setting. Settings of these kinds of costume-jewelry cameos tend to be base metal, not precious metal. Is the mounting stamped as 14-karat, 18-karat, .585 or .750 gold? While not a positive means of identification, the setting can be a strong indicator of what the "gemstone" may or may not be. Plastic and glass have a definite feel or heft (weight) to them, and even the layperson can distinguish between the feel of plastic and shell with a bit of practice. Moreover, plastic and glass are both warm to the touch and absorb body heat rapidly; shell and stone are cooler to the touch. A gemological test often used on suspect imitation stones is the hot-point needle test, in which a red-hot needle is used to put a tiny pinhole in the material in an inconspicuous spot. Plastic material will burn and produce an acrid odor or the odor of carbolic acid, camphor, formaldehyde, fruit, fish, or sour milk. Shell will not burn; only black coral will smoke, but it then smells like burnt hair.

Another way to differentiate shell from plastic or glass is by using a 10 percent solution of hydrochloric acid. Just one small drop of the liquid on the back of a shell will immediately produce effervescence if it is genuine shell. The bubbling action is common to all calcium carbonate materials, including coral. Glass and plastic do not effervesce.

Two of the best tools for distinguishing glass from shell or stone cameos are the jeweler's 10X loupe (see chapter 7 for instructions in use) and a binocular microscope. High magnification will reveal the truth about shell; it always has a straight, irregular fibrous structure. Magnification of glass will reveal flow lines, called swirl striae, and bubbles that are sometimes perfectly round but often doughnut or pear shaped. The distortion of the air bubbles in glass sometimes occurs in the cooling process.

Tool marks or the marks left around the subject that resemble shallow gouging can be seen easily in shell cameos under magnification. A quick field test that is a good indicator of the genuineness of shell is the tooth test. This can be used only on unset cameos and is done in the same way as the test used to separate cultured pearls from imitations. Gently rub the cameo against the bottom of your upper front tooth: genuine shell (as well as cultured and natural pearls) produces a slightly gritty feel; plastic and glass have a smooth, almost slippery, texture.

---

## THE ASSEMBLED CAMEOS

Somewhere between a fraud and a fake is the assembled cameo, such as the ivory bas-relief on malachite pictured in color plate 51. These cameos are not carved from one piece of material but consist of two or more cemented layers of either genuine materials or a combination of genuine, synthetic, or imitation materials. Such pieces are also called doublets. Glass cameo doublets have been made since the early nineteenth century. When the demand for cameos is high, doublets become commercially viable.

Most doublets or assembled cameos are easy to recognize. Closely examine the contact points of the cameo and its background; occasionally you can see the glue that binds the two materials. Look carefully if the background is very flat and completely uniform: the *absence* of tool marks on both the background and bas-relief will often give away doublets, both in shell and hardstone.

Inexpensive mother-of-pearl cameos, as well as abalone shell, are often sold as assembled cameos. They are prevalent in goods from Taiwan, the Phillippines, and other Pacific-rim countries.

Assembled cameos are not a modern innovation. They were discussed in the first century A.D. by Pliny: "Men have discovered how to make genuine stones of one variety into false stones of another. For example, a sardonyx can be manufactured so convincingly by sticking three gems together that the artifice cannot be detected: A black stone is taken from one species, a white from another, and a vermilion colored stone from a third, all being excellent in their own way."

### Instant Expert

#### *Points to Remember When Circa-dating Cameos*

- All artists and craftsmen were surrounded by the artistic idiom of their day and had no knowledge of later styles. They could only observe earlier styles and copy or adapt pre-existing works.
- The characteristic style of carving cameos to show color effect in the strata did not appear until the end of the first century B.C.
- Fakes and forgeries have been known since Roman times and are still being made today, especially in countries where archaeological excavations are in progress.
- Signatures in relief are usually contemporary with the gem; those in intaglio may have been later additions.
- Antique cameo subjects were generally classical or mythological. Byzantine cameos went from mythological to scriptural with Christian design.
- Cameos of the eighth to eleventh centuries are often portraits of unknown figures. There is reproduction of ancient Roman emperors and military heroes, but they are for the most part poorly crafted. Many mediocre cameos with mythological designs are found in this period.
- In the fifteenth century, there was a return to the ancient classical ideas, with the copying of old subjects and designing of new ones. Vigorous design execution with well-modeled figures marks the contemporary works as well as reproductions. Undercutting is often seen in gems from this time.
- Cameos of the late eighteenth and nineteenth centuries have strictly classical subjects because of a revival of the classical art. Fakes sold as genuine antiques from this era were often produced with subject matter unheard of in the classical repertoire, with signatures known only from old textbooks.
- Distinction can be made between cameos from the early and late nineteenth century by their subjects, as the public's interest in cameos changed: cameos were not as often collected as art as they were used as jewelry. Around 1830 the classical subjects of myths and gods gave way to romanticism and the idealized female.

- Study noses. The woman's profile with a sloped bridge and turned-up tip was conceived in nineteenth- and twentieth-century workshops.
- Lava cameos are principally from the Victorian era (1837-1901). They are found in colors from gray to greenish-brown, white, yellow, black, and beige. They may be intricately carved and usually in high relief, with subjects ranging from archaeological motifs to profiles of gods, bacchantes, satyrs, and grotesque masks.
- Leda stands next to the swan in classical art but is carved semi-reclined from the fourth century on.

*Chapter*

# 5

# MODIFIED CAMEOS AND CAMEO IMITATORS

## GLASS CAMEOS

Glass, or paste, imitations of gemstones and carved stones are nothing new. The Egyptians and citizens of Ur were using them over five thousand years ago. Later, Mycenaeans added their own skills to the glassmaking art. Few Greek glass intaglios are found from before the fourth century B.C. Glass gems were in great supply during the historian Pliny's time and furnished the poor, who could not afford real engraved stones, with the necessary signet or costume jewelry of the day. The artisans of Pliny's time imitated many stones in glass and some of these, even today, would test the skill of the gemologist. In the first milleneum A.D., glass was, because of its decorative value, considered more of a precious substance than it is today. Glass was not only scarce, making it more valuable, but also more beautiful than most gemstones, because faceting to bring out the beauty of gems was in its infancy, and deep color was the prime element of value. Glass was looked upon as a luxury, so much so that heavy taxes were placed upon glassmakers by the Emperor Diocletian, who reigned from 284 to 305 A.D.

To make the early glass signets, a mold was prepared from an intaglio by pressing it against a mixture of clay that was then carefully baked in a potter's kiln. Then a red-hot lump of glass in a soft semifusion state was gently pressed on the mold until it completely imprinted the original. If the work was done very carefully by a skillful glassmaker, the result was an almost identical replica of the original intaglio. After the piece cooled, the ragged edges and back were smoothed and polished by a wheel and abrasive powder, usually emery or corundum, after which it was ready to be set in a mounting.

In the case of cameos, the old methods are still used, combined with a few modern techniques. Once the clay model is made and dried, hard white glass can be squeezed into the hollow of the actual design, and then colored glass is pressed over it to give a contrasting background color. When the glass has cooled, it can be lifted off the mold and will show a replica of the original design. The cameo can be finished by machine or hand, depending upon the intricacy of design and purpose of the replica. A glass cameo can also be finished with a diamond-pointed tool for better articulation of design and more complete finishing.

Glass cameos can also be made by assembling different colored layers of glass, laminating them together by slight fusion or binding in some way, such as with cement glue, to produce sardonyx fakes. Glass-blowers can layer one color of glass over another for artists to carve with ordinary tools, as if the material were stone. Sometimes a combination of molding or casting the glass cameo and hand-finishing has been used.

There are some references in the old literature on paste cameos about skilled craftsmen attempting to layer white opaque glass onto natural carnelian blank stones in an effort to create a kind of layered

FIGURE 5-1.

*A large glass paste copy of a noted sardonyx cameo in Vienna depicting the family of Emperor Claudius. This copy has been identified as neoclassical A.D. 1700 to 1850, by Professor Maxwell Sommerville. (The Sommerville collection, University of Pennsylvania)*

material for use in carving cameos. The process was not very successful, and few examples are available for study.

The early paste cameos were often set with a backing of polished metal foil, which not only reflected the light through the glass, but also gave it increased brilliance and depth of color. All early artisans, from the Roman to the Celtic, used glass cameos in their jewelry. As a rule, fragments of ancient Roman glass are small and belonged to either small intaglios or small cameos; of course, a few glass cameos were of large and generous size, especially the later copies of the classical subjects (fig. 5-1).

The most common color combinations for glass cameos are white on black, green and yellow on blue, and green on brown. Now and then a single-colored glass cameo is discovered in blue, yellow, or brown glass, in both opaque and transparent materials. Some old glass displays a startling iridescence, due to having been buried in the earth.

## Cameo Glass

Besides the glass medallions and cameo jewelry, another type of ornamental glass that bears investigation comprises the cameos or bas-reliefs on cups, vases, dishes, and other vessels, called cameo glass (color plate 52).

The Roman glassmaker and lapidary were responsible for refining the art of cameo glass to its peak, producing some of the most beautiful pieces in the world. Unfortunately, only a handful of ancient cameo glass items remains intact, available for study, but those few pieces have a degree of workmanship that is mute testimony to an amazing skill.

The bas-relief carved in glass was an art form in the age of Augustan Rome. Glassblowers taught other glassmakers how to case layers of one color of glass over another. Some were made with as many as seven layers of glass. The cased glass blanks were annealed by a workman called the *vitriarius,* a man who worked with hot glass. Although the hollow, bas-relief carved vessels of sardonyx were nearly impossible to make, it was an easy task to make this type of product when the artisan was working in a fusible material. Glass also lacked the inclusions and flaws found in the natural layered stones. Exactly how the glass casing was accomplished is not clear, and the early Roman method is still not completely understood, but the following two methods may have been used:

- Dipping is a process still used today for glasses with a single casing. When blue glass is dipped into hot opaque white glass, the end or bottom of the gather (of glass) accumulates more white glass than the sides. The excess white glass is removed by tooling. This method requires a skillful operator to maintain the thickness of the outer layer.

- Using performed cylinders or cups was a method practiced in early days to layer glass. A vessel form is blown, and the glass is cased around it. The artisan must be able to produce the outer layer without distorting the already blown shape.

However it is accomplished, once the vessel is prepared, the bas-relief is carved on the top layer with engraving wheels and hand tools, such as those used for carving and scraping hardstone. You can easily identify cameo glass by running your fingers over the design; you will find that hand work has left an irregular raised relief. Polishing of most of the antique glass vessels seems to have been done with a nonrotary tool.

Four important Roman glass vessels with superb bas-relief designs can be viewed by the public. The Portland Vase, also known as the Barberini vase, in the British Museum, is believed to be the most important and beautiful piece of ancient glass in existence. This Roman vessel from the first century B.C. is translucent deep blue glass with figures carved in opaque white. The vase is 10 inches high and 22 inches in circumference with all figures cut in cameo relief. The subject is the marriage of Peleus and Thetis. The Portland Vase has no molding or assembling of the figures on the vase; all the work was done exactly as if the artist was cutting an onyx. It was produced as a deep blue vase completely covered with a layer of opaque white glass. When the glass cooled, the carver cut and drilled away the white upper layer to model the figures in the opaque white layer and cut and scraped away the excess on the dark blue body to achieve the contrasting background. The history of the vase is fascinating enough to recount. Excavated near Rome in the fifteenth century, the vase was displayed in the palace of the Barberinis until it was sold in about 1782. Eventually it was acquired by the duke of Portland, who loaned it to the British Museum in 1810. There, in 1845, it was wantonly shattered by a scene painter, William Lloyd. It has been painstakingly restored. The British Museum bought the vase in 1945, and it remains there today. It has been widely reproduced, most notably in jasper ware by Josiah Wedgwood.

The Naples Vase is an opaque white over deep blue glass two-handled amphora-like vase now in the Museo Archeologico in Naples. More heavily carved and decorated than the Portland vessel, the Naples Vase is covered with elaborate grape and ivy tendrils, with Eros picking grapes and playing the lyre for a reclining Dionysus. The carving, however, is not as finely detailed as that of the Portland Vase.

The Morgan Cup can be seen in the Corning Museum of Glass in Corning, New York. This late Hellenistic (or early Roman) handleless cup has opaque white bas-relief over a dark blue body; it was found in Turkey. The central scene depicts a Dionysian ritual. A young woman, praying for fertility, comes to make an offering before a herm of Silenus. She is accompanied by two maidens and a satyr. This cup is the earliest known fully preserved Roman cameo glass. The cup was sold in Paris in 1912 and remained in the J. P. Morgan collection until it was purchased for the Corning Museum.

The Parthian Skyphos, a Roman cameo glass cup dated from the end of the first century B.C. to the early first century A.D., is part of the J. Paul Getty Museum collection in Los Angeles. The cup is a two-handled vessel with opaque white bas-relief carvings and a deep blue glass body. It is said to be one of a pair that was found in a Parthian nobleman's tomb. Two scenes are carved on the cup: one side displays a reclining woman being served by a maidservant as a satyr observes them; on the other side, a satyr plays a lyre in front of a seated woman.

The art of cameo glassmaking was lost in the West after the fall of the Roman Empire but continued in the Eastern countries. It was revived during the Renaissance, but it was corrupted because of the imitations of ancient glass gems made and sold as antique. The outstanding European cameo glassmakers were Loetz of Austria; Emile Galle, brothers Auguste and Antonin Daum, and deVez of France; and Thomas Webb and Sons of England. George Woodall is a celebrated master of the craft; his works can be found today and are very collectible. Cameo glass has never been produced to any extent in the United States, but there has always been a steady supply of collectible pieces.

---

## TASSIE GLASS

In the past much secrecy surrounded the making of glass, because it was a far more precious commodity than it is today. In the eighteenth century, cameos and intaglios of ancient times were extensively copied in glass by James Tassie, a young Scotsman from Pollockshaws, near Glasgow. Originally a stonemason, Tassie had outstanding skills and achieved great success as a craftsman of cameos and engraved gems. Tassie made history by developing a special formula for making glass. Although Tassie made a big show of maintaining the secrecy of his recipe, there was actually no real secret to the ingredients or process for the glass that he used to imitate antique cameos and intaglios. The secret composition of the glass was a finely powdered potash-lead glass, opaque white and colored, which was softened by heating and pressed into plaster of paris molds. Tassie encouraged the aura of mystery solely to prevent others from entering what proved for him to be a lucrative field.

Tassie lived at a time when cameos were popular, and the status-conscious consumer kept the trend of wearing them in jewelry alive. The many people who could not afford the genuine article in hardstone were delighted to settle for the cheap glass substitutes Tassie provided. His skillfully made colorful imitations were suitable for art cabinets as well as for the more conventionally mounted jewelry. The replica collection numbered over fifteen thousand, an especially large number because he had managed to wrangle permission to duplicate some of the outstanding cameo cabinets in Europe. In 1775 Tassie published a catalog, *Reproduction of Gems Ancient and Modern,* illustrating his models; the catalog itself is now a much-sought collector's item.

During his most productive years, Tassie turned out thousands of Lucretias, Cleopatras, Minervas, cupids, nymphs, and satyrs. On one order alone, commissioned by Catherine the Great of Russia, his entire range of designs—thousands of models, ancient and modern—were made for her, in triplicate.

The portraits on the Tassie cameos are generally contemporary aristocrats and royalty. Some of the pieces are marked with a *T,* or "Tassie F." After his death, a nephew, William, inherited the business and continued expanding the collection of replicas. William's production is marked "W Tassie F." The business thrived well into the nineteenth century but faded after William Tassie died.

## WEDGWOOD

James Tassie was not the only renowned glassmaker in the eighteenth century. The English potter Josiah Wedgwood (1730 to 1795) perfected a technique of producing cameos in a hard fine glass he called basalt, or Egyptian black; then in 1774 he developed the jasper ware, which is still familiar today.

The pottery firm was known as Wedgwood and Bentley in the 1770s. Their cameos were made from a white porcelain bisque and unglazed jasper ware, which had a white relief on a colored or black background. The pieces were meant for art-cabinet collections and decorative use on furniture. A catalog of Wedgwood designs published in 1787 listed 1,764 different cameos and 400 intaglios, with subjects ranging from the classical Greek and Roman to portraits of philosophers, poets, and heroes. Some of the subjects that appear in cameo were also made in intaglio, often on the reverse of the same stone. Wedgwood was so highly regarded by his contemporaries that kings, princes, and collectors from all over Europe allowed him to study and copy (take molds from) their collections. Designs to be found on the Wedgwood cameos include masks, portraits of ancient as well as contemporary men and women, and classical Greek and Roman subjects (color plate 53). He also took designs from medals, coins, wall paintings, fresco bas-reliefs, sarcophagi, and engravings in books. The cameos were ambitious in size, ranging from the size of a bean to over 54 inches. Some cameos were grooved on their edges for mounting in furniture; others were pieced so they could be sewed onto fabrics; and still others had holes at the center so they could be used on strands of beads.

Largest of the cameos are the medallions. Mounted to use as ornament, they are called "Applied Wedgwood." The small round ones vary from 2 to 5 inches; plaques in oval, round, or octagonal shapes, from 5 to 11 inches; tablets and rectangular-shaped bas-reliefs, from 5 to 54 inches. Medallions were used as decoration in boxes, trays, cabinets, tables, pianos, desks, tea caddies; they were framed for wall furniture and sometimes used most effectively as an architectural treatment over a doorway, fireplace, or window. The famous Amethyst Room, designed for Empress Catherine the Great, had Wedgwood cameos set in chimney pieces and walls as part of the room treatment. Thomas Jefferson had the mantelpiece in the dining room at Monticello inset with Wedgwood cameo plaques.

Small Wedgwood stones for jewelry began as intaglios, which could be used for sealing letters but were also worn by women as costume accessories. The little seals were usually of polished black basalt, and so it was natural that the first jewelry cameos made by Wedgwood were from the black basalts. He began by buying model cameo examples in plaster and molds from James Tassie. Then, after he perfected jasper ware, that material quickly became the preferred material for cameos. Jasper ware can be stained with various metallic oxides in many colors to produce an attractive product. Jasper ware and jasper dip (white inside) were made in various shades of blue and sage green. The rarest and most valuable color is crimson, then yellow, olive green, and dark green. Blue and sage green are most common.

Cameos are produced by assembling the white opaque bas-reliefs onto a jasper ware blank and then firing it into one piece. The firing produces a hard, fine-grained, unglazed, slightly translucent stoneware cameo. Although it is generally left with a matte finish, jasper ware is capable of taking a high polish.

There are numerous books on the life and production of Wedgwood, many of which delve into his involvement in social and political issues of his time. His involvement in social affairs, his many acquaintances with royalty, and patronage from the European aristocracy enabled him to develop design ideas and continue to gather accolades for his art.

A Wedgwood cameo button is a great and rare collectible, usually the prize of any collection. Josiah Wedgwood, ever with an eye for good business, made the following comment about cameos in his 1787 catalog: "The cameos are employed for various ornamental purposes . . . as also for buttons, which have lately been much worn by the nobility in different parts of Europe." The buttons come in round, oval, curved, and rectangular shapes; they range in size from ¼ to 2½ inches, to accommodate both coats and waistcoats. Buttons are framed in a wide variety of materials: gold, brass, cut steel, marcasite, and mother-of-pearl. An authentic button will have "Wedgwood" imprinted into its back.

Collectors enchanted by the Wedgwood models and designs are aided in their search for additions to their collections by the fact that most genuine specimens are easily recognized. As early as 1759, Wedgwood began to put his name on his wares, and from around 1769 the goods were stamped with the names of the partners, "Wedgwood and Bentley," in a circle pattern, until Bentley's death in 1778.

In what has been a good move for collectors, the company began impressing its name with a metal stamp into the backs of the clay bodies of cameos. The impression is purposely difficult to remove, and the use of acid or a grinding wheel to force removal leaves a depression in the stone.

The simple name "Wedgwood" was used continuously until 1891, after which the word "England" was added to comply with the McKinley tariff laws in the United States, which required all imports to be marked with the country of origin.

Collectors should not simply assume a piece is Wedgwood because of its markings, however; they must inspect the name carefully and recognize the proper type style. The overwhelming popularity of Wedgwood has made copying by potters all over Europe and the United States common.

## SULPHIDES

Cameos are also incorporated into decorative objects with the cameo incrustation process. This treatment consists of taking cameos of any subject, usually portraits, and enclosing them in glass, so they become chemically imperishable.

The art of cameo incrustation, called sulphide, was invented in 1750 by a Bohemian, whose first attempts were not very successful. However, they piqued the interest of some French manufacturers, who later invested enough time and money into the process to incrust numerous medallions of Napoléon. They then sold the sulphides for enormous sums of money and launched a profitable business.

In the first patent, the process was called *crystalloceramic,* and specimens of the incrustations have been shown in public exhibitions in decanters, wineglasses, lamps, paperweights, plates, scent bottles, tumblers, medals, and doorknobs.

The bas-relief bust or portrait is entirely isolated in a coating of white flint glass. The incrusted cameo must be made of materials that require a higher degree of heat for their fusion than the glass; usually, a mixture of china clay and potash is used in such proportions that it will harmonize with the density of the glass. The desirable result of this process is that the encased cameo appears silvery or metallic, due to the refraction of the light through the glass.

## PLASTIC CAMEOS

Plastic is any of a number of resinlike synthetic substances, which can be easily molded by heat, pressure, or both. In the last few decades, with the development of new and better technology, cameos have been made of plastic in increasing numbers for use as costume jewelry.

An exceptionally wide range of colors can be achieved with plastics, including the contrasting colors needed for cameos, especially for imitating shell or Wedgwood cameos. If examination of these imitations reveals no separation planes between the figures and their backgrounds under magnification, they may not have been assembled from separate components but molded in a two-step process, with the figure (subject) portion of the mold being filled first, followed by pouring of the background material. If the figure has not been allowed to set properly before the background is poured, you will see a swirling of the two colors around the subject.

Distinguishing between plastic and genuine shell or hardstone should not prove a problem; tips for making the distinction are given in the previous chapter. The following points should be remembered:

- Plastic materials are always warm to the touch; shell and hardstone will be cooler when first handled.
- A high or waxy luster or lack of sharp outline of the figures should arouse suspicion of plastic.
- Plastic cameos are generally very inexpensive, while shell and hardstone can be $50 and more.
- When tapped with a fingernail, plastic usually produces a dull thwack, not the clear sound made by shell and hardstone.
- Plastic has a synthetic look to it and a molded appearance.

---

## CAMEO WAX PORTRAITS

As early as the fourth century B.C., the ancient Greeks, Assyrians, and Egyptians were modeling figures in wax, or ceroplastics, as it is sometimes called. While little is known of their tools, they probably used their fingers and thumbs, with detailing tools made from bone or ivory.

Many studios and glyptic arts carvers enjoyed royal patronage or worked for aristocratic collectors in Italy, France, Germany, and England. The early artisans also worked in wax, and cameos became a separate and important field of collecting. When the sumptuary laws—royal orders concerning which gems, jewelry, clothing, and colors could be worn by whom—were relaxed, the desire for cameos spread to the average man and woman; and for those who could not afford the genuine materials, waxes became a pleasant substitute. In the late 1700 wax models of Tassie's 15,800-piece collection were mass-produced to fill the market demand. This type of collection was for study and pleasurable viewing but not for wear.

A wax model was often the preliminary to a more formal portrait carved in stone, etched in metal, or painted, and it was produced on a modeling board. Occasionally the modeling board also became the background for the finished model. But often wax was the intended medium for the finished portrait. Wax was extensively used by early sculptors, goldsmiths, and artists to model their subjects, in much the same way a painter makes a sketch of a proposed larger canvas. Sometimes a wax-model portrait was used in the lost wax method of casting in opaque white glass; the resulting bas-relief portrait was then bonded to a colored glass blank to complete the assembled cameo for use in jewelry.

The finest waxes ever produced were those of the Italian Renaissance. Indeed, the sixteenth century was impressive in its talent. The famous goldsmith and sculptor Benvenuto Cellini said in a treatise on art that Pope Leo X and Francis I of France vied to see who could gather the greatest artistic talent around him. Of the great engravers, sculptors, goldsmiths, and carvers working in wax in the sixteenth century, the most widely recognized name is that of Cellini himself. Cellini has been credited with spending over two hundred hours on waxes of cardinals' medals, and several of the artist's pieces are preserved in Roman museums. Of particular note are the cardinals' seals, which Cellini especially enjoyed creating. He recounts in his autobiography the designs of two seals produced first in wax, then in metal. Both seals were about 8 inches in length and designed in an almond shape; each bore the individual cardinal's title on it in the form of a rebus. (A rebus is a representation of words or syllables by pictures of objects or by symbols, whose names resemble the intended words or syllables when spoken aloud.) Cellini made a seal for the cardinal of Mantua engraved with the ascension of the Madonna and the Twelve Apostles, along with the rebus conforming to the cardinal's title. Because Cardinal Ippolito of Ferrara had a twofold title, his seal was split down the center, with one side depicting Saint Ambrose on horseback with a whip in his hand chastising the Arians, and the other side showing Saint John the Baptist preaching in the desert.

Leone Leoni, a carver of note, presented a portrait in wax of Michelangelo to the master Michelangelo, who promptly returned the compliment by giving Leoni a wax creation he had made of Hercules.

From the middle of the fifteenth century into the sixteenth, wax modeling was a highly regarded art in its own right. The portraits made in colored waxes and finished in detail with the application of precious stones to the subject became known as jeweled waxes.

In vogue at different times in Germany, France, England, and the United States, wax cameo portraits progressed from good to exquisite during the seventeenth and eighteenth centuries. In France they were also popular as death masks. By the early nineteenth century, the trend for wax cameos had run its course, and with the invention of the daguerreotype, wax portraiture as a popular art form slowly faded away.

Although this type of cameo art is not well known, a number of antique wax portraits can be found in public and private collections. An unusually fine wax portrait of the duke of Wellington by Thomas Wyon, chief engraver of British seals in 1816, and a bas-relief wax portrait of Empress Catherine II of Russia, circa 1779, are part of a private collection in Florida (color plates 54 and 55). A large public collection of wax portraiture is on display in the Philadelphia Museum of Art.

## PLASTER OF PARIS AND SULPHUR CASTS

The craft of making impressions from carved and engraved gemstones reached its peak in the eighteenth century. Cameos cast in sulphur or gesso (also called impronte) impressions were a part of every major collection of church and state (fig. 5-2). They were important tools for study and even more important as verification of the original engraved gems. Their existence sheds light on the intellectual curiosity of the era, how learning flourished even in the home of the common man.

The casts were not easily won, however. There was argument and dissension between some collectors with full cabinets of engraved gems and the artists who wished to replicate them, even when the artists were celebrities in their field such as Giuseppe Torricelli and Bartolomeo Paoletti. The director of the Uffizi in Florence argued to the grand duke of Tuscany, Ferdinand III, that by making the casts, the fame and value of the royal cabinets would be expanded. The grand dukes had theretofore been

FIGURE 5-2.

*A plaster of paris cast of Cupid in his cart (From the collection of The Lizzadro Museum of Lapidary Art, Elmhurst, IL)*

reluctant to allow any replication of the engraved gems, citing the danger of damage to the stones along with an undesirable flood of cast duplicates. Paoletti won the argument but was allowed to work only under strict supervision. He was limited to making a few casts per day and absolutely forbidden to duplicate matrices or keep those casts that were imperfect. Much of what we know about this project is preserved in letters in the archives of the Uffizi.

The casts are rare and of great scholastic interest. Many of the 7,189 Paoletti casts are displayed in the Museo di Roma. Occasionally a few casts surface for sale in the United States, such as the collection of twenty-seven cameo casts from the P. Paoletti atelier offered by a dealer in Connecticut (fig. 5-3). Treasures such as the three-case collection pictured in fig 5-4

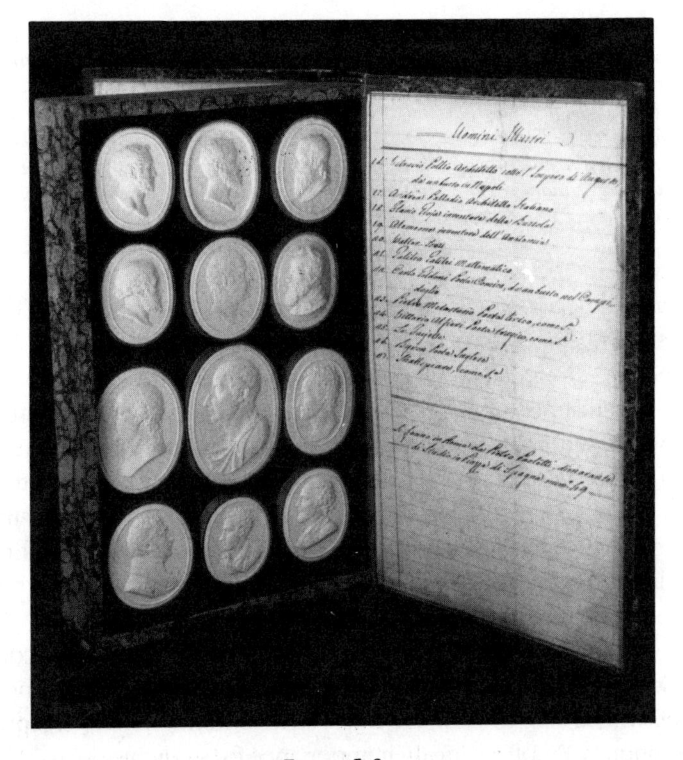


FIGURE 5-3.

*Twenty-seven cameo casts, each encircled with a gilt frame, set against a blue background in a custom-designed case. The maker identified himself as P. Paoletti, with a studio at 49 Piazza di Spagna in Rome, and wrote descriptions of the cameos by hand. The subject matter includes gods and goddesses, portraits of illustrious figures from the fields of architecture and English and Italian literature, and others such as Galileo and Lafayette. (Photo courtesy of The Book Block, Cos Cob, CT)*

FIGURE 5-4.

*A grand display of cameo casts in three volumes that are actually cabinets for the display of the cameos. Listed on the covers are the titles of the pieces displayed. Cabinets like these were popular in Rome in the late 1700s and early 1800s. Stampings on the spines suggest these casts are the work of the firm Liberotti. The casts depict cameos of the classical sculpture of Greece and Rome as well as architectural details from ancient public buildings, private palaces, and churches. There are also casts taken from subjects of paintings by Titian, Giovanni da Bologna, and Leonardo. The volume on the left contains nine portraits of philosophers. (Photo courtesy of The Book Block, Cos Cob, CT)*

are rare offers; not only are they in perfect condition, but they are housed in custom-designed display cases with handwritten manuscripts describing them.

That cameo casts remain in good condition over the centuries is another argument in their favor. The hardness of the materials used for casting preserves the pieces without deterioration, and because they are casts and not originals, many have survived without the mutilation so common to original works of art. Also, because they were meant for study and display and not for wear, they were not changed or abraded by daily use, nor corroded by long entombment in a burial place.

Researchers of this subject point out that the cameo collector should take every opportunity to study collections of casts. By doing so he or she will be able to compare many styles and develop a perception of true antiquity. Emphasizing their importance for study, historian C.W. King declared, "(Gaining the ability) never to pass over an antique for a modern work is a faculty acquired in this way." Further, he promised that with careful study of casts, the actual type of gem that produced the impression could be distinguished, not just the school to which the style belonged.

Recipes in ancient texts on plaster of paris casts recommend that if one wishes to eliminate the glaring whiteness of the piece once it is cast, dipping it in skim milk will give the appearance of marble; a dip in strong tea will produce a tint of brown. A durable cast can be made in sulphur and colored with vermilion, melted, and poured over the gem before the molten sulphur is added. The advantage of this method, it has been noted, is the superior hardness over plaster of paris.

*C h a p t e r*

# 6

# APPRECIATING CAMEOS AS ART

To appreciate cameos as art, one must comprehend good design and recognize excellent workmanship. Cameos are judged as art using some of the same criteria as other fine arts: beauty and desirability. Should cameo appreciation and value spring from an emotional reaction, an assessment of the expression and vitality of the piece, or should cameos be viewed using strictly intellectual criteria to judge specific art qualities, such as craftsmanship? The answer, naturally, lies somewhere in between.

Art of every form conveys an emotional message, which is the motivating force of a work. Artists use line, tone, shape, pattern, texture, color, space, movement, and volume to convey a message. In the very small space of a cameo or intaglio, the artist is severely limited as to what he can and cannot do. Therefore, to appreciate this art form, one must be aware of how well the artist has met the demands of the medium and how well he has conveyed his message by his craftsmanship.

While a vast quantity of cameos made today are mass-produced clones that can scarcely be considered art, a few craftsmen still tend the flame of artistic creativity. But even the most passionate and creative artists often have divergent views of art and of what is, and what is not, beautiful. There are cameos carved in hardstone and cameos carved from organic materials such as shell; each has different emotional overtones. Because cameos are representational art—that is, art that represents facts or creates a picture of something real or imagined—they can also portray emotions such as grief or love and qualities such as strength or patience. Moreover, representational artists can express depth and distance within the confines of a very small space, as long as they observe the principles of composition in the use of proportion, perspective, light, and shade.

There is a persistent notion that the art value of cameos can be judged by a stable criterion, with the most unrelenting value element being the design. This is true, because good cameo art must have an orderly composition; without it the image will be confused and chaotic. How is an orderly composition produced? It begins with an artist (the carver) blocking out the design so that the main lines have a relationship to each other. Artists in all media recognize this principle. A well-made piece will have an organized pattern; a pyramidal, circular, or sweeping arrangement of lines to balance the design and focus interest (color plate 56). The pyramidal (triangular) pattern is the most effective in directing focus. The center of the triangle is a natural focal point of the design. The sides of the pictures contain the details, and the base unifies the entire scene.

A combination of graphic elements and representational elements often was used in old cameos. Today's carvers are not as likely to incorporate graphic elements, such as patterned borders, opting instead for a simple, framelike line border around a picture.

Few things are as subjective as the opinion of what constitutes good art; moreover, opinions change over time with cycles of taste. Nevertheless, one must be willing to recognize some fixed qualities of excellence — such as harmony, balance, and rhythm — if the superior cameo is to be distinguished from the mediocre, so that cameos, especially contemporary pieces, that are artistically first rate and thus have greater appreciable value, can be selected.

*Harmony* in design simply means that the lines and shapes in the picture have something in common, so that when they are combined they agree with each other. For example, it is more pleasing to the eye to see a combination of circles in various sizes or triangles in various sizes than to see the circles and triangles in one piece. Harmony implies likeness, and it is closely associated with balance. *Balance* is the arrangement of shape(s) to bring equilibrium to a scene (color plate 57). That is, a design is divided so that the halves are identical in shape and position, balanced and symmetrical. *Rhythm* relates to the impression of action or movement. Color plate 58, *The Letter,* carved in shell, shows rhythm in several ways: a billowing curtain caught by a gentle breeze; a pigeon understood by the viewer as having just swooped through the open window with a letter; and a rider waving from the road. The artist creates rhythm by depicting swirling garments or draperies, spiraling smoke, or a vital posture or pose.

Yet another element is used to judge a cameo's art quality and assess its monetary value: style. *Style* is derived from the Latin word *stilus,* a writing tool, and it originally referred to a distinctive way of writing. Today, however, style means the way in which something is done; in art, style means the particular way in which a work is created. To art historians, an artist's style gives clues that enable them to determine where and by whom a given work was made and the period and setting in which the artist worked.

Certain recognizable characteristics mark an artisan or craftsman, period, or culture, and these constitute style. Style is revealed in the form of the creation. In various periods people have perceived the world differently; for instance, the naked body has been deemed splendid in some eras, shameful in others. The integration of subject matter, culture, and norms of a society constitutes style; for example, it may be possible to find a nude Adam on a cameo, but do not expect a nude Caesar.

Just as the professional art critic analyzes many separate elements of a work—even one as small as a cameo—to understand the entire piece, so should the serious collector.

Combining all the aforementioned fundamentals and using them to evaluate a cameo's art quality or monetary value may seem like investigation overkill or, at least, too much trouble to the average collector. But when considerable sums of money often rest on the authentication and

**FIGURE 6-1.**

*A sardonyx cameo bust of Augustus wearing the aegis, about 1½ by 1 inch (3.7 by 2.8 centimeters). (The Metropolitan Museum of Art, Purchase, 1942, Joseph Pulitzer bequest, 42.11.30)*

provenance of an antique cameo, extraordinary caution and time-consuming study are sometimes necessary. Art historian-appraisers are cognizant of how small subtleties in style impact on circa dating and affect value. Some professional appraisals are so detailed in their investigation and explanation of a specimen that every minute detail is scrutinized. The following report by Ellen J. Epstein, a New York art historian-appraiser, provides a vivid illustration of authentication. The evaluation concerns the sardonyx cameo of Augustus in fig. 6-1, dated as first century A.D. The appraiser does not believe the cameo is as old as purported and supports the view with the following scholarly reasoning and research.

### *Appraisal of the Augustus Cameo*

*The Metropolitan Museum's sardonyx cameo of opaque white on a rich brown ground shows Augustus portrayed in glyptic form, with an aegis over his left shoulder, a laurel wreath in his hair, and a spear over his left shoulder.*

*G. M. A. Richter,* Engraved Gems of the Romans, *described the cameo as "a famous piece with a long history." Whether its history is actually long is questionable.*

*It was first known in the Arundel Collection, originally formed during the reign of Charles I (1625-1649), then passed into the Marlborough Collection, then into the collection of Arthur J. Evans; and ultimately, in 1942, it was purchased by the Metropolitan Museum, together with thirty-one gems from the Evans Collection.*

*Augustus's face and neck are seen in profile, although the rest of the bust is shown in an unusual, but not unique, reflection of the larger Strozzi-Blacas cameo in the British Museum. However, the Metropolitan cameo includes the right shoulder and upper arm, incorporates a larger bust area, and is more clearly a back view.*

*On the whole, the state of preservation of the Metropolitan cameo is very good. There is virtually no evidence of damage in the background portion. There are indications of wear and use on the face; e.g., the scratches around the cheek area, the ends of the fillet and spear are damaged, the nose and ear are broken. Noteworthy is the contrast created between the articulated volume of the milky white portrait and the flat dark background, which in its own terms is effective and impressive.*

*The profile physiognomy agrees iconographically with that established for Augustus: The furrow of the brow, slightly hooked nose, broad forehead, protruding chin, Venus rings—the locks of the forehead hair and those of the nape of the neck—wide-open eyes, and the shape of the lips. Richter recognizes the wide-open eyes as suggesting the "intensity of his (Augustus's) gaze, which according to Suetonius, greatly impressed his contemporaries." However, in comparison with the Strozzi-Blacas cameo and coin portraits, the eye representation of the Metropolitan gem appears exaggerated and somewhat outside of the canonical iconography of the Augustan age. The same situation may be noted for the overly arched upper lip, which echoes a feature that becomes fashionable and thus emphasized in a post-Renaissance era.*

*Both Furtwängler and Richter have observed Augustus is represented as older and more mature on the Metropolitan gem than he is on the Strozzi-Blacas gem, thus "resembling the Prima Porta statue and the portrait on the Vienna cameo." Furtwängler also noted Augustus's disturbed facial expression and ardently energetic glance. The mouth, nose, eye, cheek region, the throat and wrinkles of the brow are, in Furtwängler's description, "carefully expressed and modeled with a sense of liveliness." These observations have led him to establish the gem as a work worthy of the master gem craftsman Dioskourides. The feeling of tranquillity conveyed by most other portraits of Augustus is here reinterpreted. Augustus now appears to be within a realm that is totally immersed in the vicissitudes of his imperial role. The resulting intensity is notably foreign to other portraits of Augustus, but this in itself, though an enigmatic feature, cannot serve as a basis on which to question a first century A.D. date.*

*The laurel wreath is an old device appearing on representations of Hellenistic rulers and gods and, in the preceding Roman period, on coins of Julius Caesar. As it is used by Augustus, it implies not so much imperial or majestic but acquires a different interpretation as an ornament for festive occasions or as a military insignia. Later in the first century, it appears only on representations of the imperial family, and again is reinterpreted. The laurel wreath does not appear on the Strozzi-Blacas gem, and since the now extant diadem is an addition of the Middle Ages, it is not known what the original fillet was like. What is puzzling about the wreath and fillet of the Metropolitan gem is not then, its presence, but its enigmatic form. The fillet terminates at the top of the head in a most disturbing and unsatisfactory fashion, with a featherlike top that is both unique and unconvincing. No analogous portrait manufactured in antiquity can be cited which compares with this. It seems quite inconceivable that an ancient artist would produce such a frivolous construction. In addition, the distinction between the locks of hair and the leaves of the wreath is merely suggested and is barely distinguishable without magnification. This treatment is alien to other Augustan comparisons clearly observed on Augustan coin portraits, where hair and leaves are precisely delineated and textures are well defined. It hardly seems likely that an Augustan craftsman was responsible for this type of rendering.*

*The opposite end of the fillet-wreath, rather than terminating in a natural, gravitational fall—as all representations of fillet tend to do in almost all ancient periods—seems instead to possess an independent life of its own and whimsically blows outward with a decorative sweep. The entire treatment is more reminiscent of the windblown effect seen so frequently in the works of artists such as the seventeenth-century Poussin. The introduction of this additional element of reinterpretation of the Roman background area is perhaps one of the most telling details which support a much later dating, later than Dioskourides.*

*The motif of the aegis with gorgonion included in a portrait can be traced to Hellenistic representations. The addition of a second head—that of a bearded man with wings—has precedence, notably in the Strozzi-Blacas gem and the Gonzaga double portrait cameo in Leningrad (which may be dated as early as the third century B.C.). The winged figure on the Metropolitan cameo, in relation to the aegis, can be regarded as a reflection of the Strozzi-Blacas gem. It does not seem superfluous to conclude, particularly noting the similar placement of this figure, that the artist of the Metropolitan gem was at least familiar with the Strozzi-Blacas motif, if not with the gem itself.*

*There are several factors that can be read as a misinterpretation of the earlier piece. It is, for example, a misunderstanding and lack of identification with the classical spirit that could produce the placement of the bearded head in such an unbalanced position as it is on the Metropolitan cameo. The cropping of the beard is a total misconception of what an ancient artist would do. The head is shown as resting against the shoulder of Augustus, with the left wing extending beyond the aegis, onto the bare skin—another unusual technique—echoed in the extension of one of the snake heads beyond the border of the aegis onto the bare back area, allowing the smooth surface of the flesh to serve as a background for the well-defined outline of the snake head. This approach to an ancient device seems quite irregular, and to this point, no counterparts can be cited.*

*A comparison of the surface treatment of the fabric of the aegis leads again to an uncompromising position. The artist of the Metropolitan gem has translated the aegis material into a scaly, almost fishlike surface, which is unparalleled in other aegis representation, though the conception is not far removed from the Gonzaga gem. The portion of the aegis that opens to reveal the left shoulder is suspiciously akin to the corresponding area of the Strozzi-Blacas gem, which by its lack of clarity leaves the area undefined.*

*On the Metropolitan gem, Augustus's back is not completely turned toward the viewer; his left shoulder is lower than the right, perhaps forced down by the weight of the aegis, and the left shoulder blade is given in high relief. The tension created seems to have been intentionally avoided by ancient gem craftsmen, who preferred more static poses. The upper right arm is unnaturally affixed to the shoulder and side, a detail that does not find a place in analogous ancient representations. The large bust area is unusual in Augustan times and further indication of a later date. The volume of the lower area, which unsatisfactorily overwhelms the portrait area, is also suggestive of more modern workmanship.*

*While it is difficult to see in photographs, a firsthand examination immediately reveals a strange portrayal of the ear area. The circular, deeply cut incision of the ear is strikingly disturbing, and the viewer, at first glance, is made aware of its existence. The area surrounding the ear is awkwardly rendered and only by turning the gem on its side can one discover the extent and shape of the ear itself. What becomes clear with detailed examination is a curious merging of the ear with either one of the laurel leaves or with one of the bows of the fillet. The ear itself is not in as high relief as might be expected, and the whole area of the ear, wreath, and fillet is awkward. The lack of overall precision which distinguishes fine ancient craftsmanship can only add to the question of authenticity of this gem.*

*The gem is one of baroque decorativeness that is unfamiliar in the sober Augustan age and more properly belongs to an age of neoclassicism; nineteenth century seems stylistically acceptable.*

It is not suggested that this type of exhaustive review of an antique be conducted on all cameos, nor is it necessary to evaluate modern ones stylistically. Rather, it serves as a model of in-depth investigation, diligence in authentication, and logical reasoning in how a cameo can be dated based on stylistic grounds.

What the foregoing demonstrates is that many interpretations of ancient cameos exist; the experts do not agree; and much research remains to be done in this field. The value message is clear, though. If a cameo is authentic and dates from the period of its style, value is greater. If it is only in the style of a period, value will be less, because it will be perceived as a copy.

## ELEMENTS OF VALUE

Value is culturally defined. The value of an object in one culture or society may not be the same in another time or place. This is as true today with jewelry and art as it was in the past. When a society changes, what was once considered valuable may become useless. Many staple items once considered desirable and valuable—from button hooks to buggy whips—become obsolete in industrial and technological revolutions. This happens more often with utilitarian objects than with art objects, which, generally at least, have intrinsic or aesthetic value. So it has been with cameos over the centuries. Fortunately, many cameos can be classified as art and have at least aesthetic value. Those carved on precious stones also have intrinsic value, the market value of the gem material itself.

In determining the monetary value of a cameo (but not archaic Greek and Roman, which are judged by different criteria, explained later in this section) there are some fundamental elements to consider. Several are the same as those used to judge artistic qualities, and all are factors a qualified jewelry appraiser would consider in estimating market value: design, craftsmanship, subject, material and its quality, condition, authenticity, and signature. Rarity and age, elements tied to the laws of supply and demand, are used to estimate the value of archaic or ancient cameos, and by themselves may be the sole reason an item is deemed valuable.

How to extract the most from these elements, give them a value, and reach a documentable market price is the work and art of the jewelry appraiser and is usually foreign territory to the collector.

Design, discussed earlier, remains the primary factor in establishing the monetary value of a cameo. There should be excellent design, even in the ultrasonic clones, which have lower values simply because they can be repeated in endless quantities. In his *Standard Catalog of Gems,* John Sinkankas mentioned that if one is not sure how to judge the artistic merits of an engraved gem, remember that "the finest workmanship is normally found on the best materials by the most skilled engravers producing the most artistic designs possible."

Craftsmanship is a particular guidepost for the cameo appraiser. The inferior cameo is poorly carved. Details such as eyes, noses, and mouths in portraits and leaves and flowers in floral arrangements should be carefully inspected. Poor work will often have shaping defects: sausagelike arms and legs; disproportional arms and legs; a nearly featureless face; weakly cut grooves to depict hair or other design features. A well-contoured face and body on a cameo denotes a first-class craftsman. The superior work shows crisp edges without any slips of the engraving tool. A well-cut cameo will stand up to close inspection with a magnifying loupe, while flimsy craftsmanship inevitably shows modeling faults that are visible even to the unaided eye.

In German workshops today it is popular to carve in a very thin layer of the white strata in low relief (color plate 59). The Idar-Obserstein master carvers are near genius in manipulating the thin layers to achieve some dramatic illusions. They can project the subject to look as if it rises much higher above its background than is actually the case. The successful low-relief carving creates the effect of depth while maintaining width and height. Although increased monetary value has normally been assigned to the deeply carved cameo and those in high bas-relief, this may be changing due to the German proclivity for low-relief carvings.

The subject matter of a cameo may be specific to the creator of the piece, but some subjects simply sell better than others and are more valuable. In the past jewelry appraisers have given more value to cameos with full scenes and many figures; next, cameos with two or more men's or women's portrait heads on a single cameo; then, the single portrait. The order of difficulty in carving proceeds from the nude human figure to portraits, draped human figures, animals, plants, and objects. Some historic or legendary scenes, as well as portraits, are more in demand than others and are thus more valuable. Ultrasonically produced cameos with the profile of a woman are inexpensive. The shell cameo hand-carved with a woman's profile varies in price according to craftsmanship, but the majority of the contemporary pieces are not costly. To become proficient in judging quality in modern shell cameos, make a firsthand examination of them as often as possible, and do as professional jewelry appraisers do: build and use a quality rating system, not only to assess value but as a visual aid in building a personal collection. The way to create this system is to purchase three cameos that you can quality-judge as good, better, and best. These cameos are the controls and references on which a value basis is made and an understanding of the nuances of quality arises. The three shell cameos in figures 6-2, 6-3, and 6-4, from the Sebastanelli firm,

**FIGURE 6-2.**

*Example of a cameo that would be rated as* good *on a rating scale.*

FIGURE 6-3.

*A cameo that would be rated as* better. *Notice the additional devices and improvement of artistry in the subject.*

FIGURE 6-4.

*A cameo that would be rated as* best *on a rating scale. The fine workmanship and skill involved in making this cameo is readily apparent.*

demonstrate the concept very clearly. Note the progression of detail among the cameos, the strength of improvement, and the introduction of additional elements in the design as the quality of the cameo improves from good to best. One factor to keep in mind when assessing modern shell cameos is that the more devices, such as flowers, birds, jewelry, and so forth, that are displayed, the more expensive the piece. This is explained in the most basic terms in the Sebastanelli workshop: "More work, higher cost." Sandro Sebastanelli also noted that the best cameo is five times more expensive than the good model.

The cameo's finish is produced by polishing. Though a high polish may mean the carver has expended a lot of labor, it does not automatically mean the work is exemplary. In Asian countries, for example, the carvings of nephrite jade and rose quartz often have a very high polish to divert attention from the many faults in the carving. The same ploy can be used on shell and stone cameos. Some carvers believe that the human figure requires various finishes: a soft finish on the face and hands, with hair and clothing polished or polish highlighted. It is wise for the cameo collector to remember that even though a high polish may bring out the natural beauty of the hardstone, it may not necessarily enhance the carving itself.

Hardstone material is judged by its color, pattern, texture, and transparency. It must be determined if the material has been dyed, heat-treated or color-enhanced, whether it is natural, synthetic, assembled, glass, or plastic; all bear on the value. Maximizing the best qualities of the material is essential to good craftsmanship; an example is the bloodstone cameo in color plate 26. A judgment must be made as to whether or not the carver has taken full advantage of any regular or irregular features of the material. And, of course, in any monetary evaluation, not only the use of the material but the type and quality of stone play an important role. Clearly, cameos carved in emerald, ruby, and diamond are going to be worth more than those carved in sardonyx; the quality of these stones will help the collector and appraiser estimate just how much more.

Condition is emphasized by the knowledgeable collector, and rightfully. Cameos that have withstood the ravages of time and survived relatively intact are preferred to fragments or damaged pieces. The damaged cameo is acceptable only if it is documented as a rare antique. Shell cameos are especially vulnerable to cracking and should be carefully examined by being held up to the light. Vertical lines, faint to severe, indicate cracks. Repair is impossible, and such cracks will reduce the value of the cameo substantially; the amount of the reduction will be proportional to the nature and predictability of the crack.

Authenticity is closely aligned with provenance, or origin: what is the piece's history, and who owned it? The question of authenticity is interesting and often confusing; it can significantly affect the value of a cameo. No matter the product, it always seems to have more value if a well-known individual made or owned it. To prove authenticity of a cameo often requires a lot of research and deductive reasoning. Some collectors search for and buy additions to their cabinets on the strength of authenticity alone; they judge a cameo by its provenance and remain unimpressed with one that may be considered beautiful but whose provenance cannot be established. Even museum officials are skittish about acquisitions based solely upon an item's beauty and will choose an inferior example of a known master carver when the work can be proven genuine.

Authenticity and provenance can sometimes be established by family records or sales receipts. Some antique cameos may have documents if they have been deaccessioned from a museum collection; however, the papers may be difficult to obtain. Museums seldom give papers to purchasers of deaccessioned items; in fact, it may be the policy of the museum not to trade or sell items from collections, so the institution is not anxious to have any sale publicly known. Numbers painted on the back of a cameo are a telltale sign of museum record keeping; they indicate that some documented provenance exists someplace. Items from famous private collections may be found illustrated in early glyptic books. And, in the last fifty years, scores of auction catalogs (helpful too for price watch and quality comparisons), museum exhibition catalogs, and dealer catalogs have been published. They frequently repeat the illustrations of both rare and common cameos.

Any signature on a cameo demands careful inspection and validation before assigning a value beyond the normal market price. Signatures are the weakest links in the chain when attempting to authenticate a work. The names of celebrated contemporary carvers are still easily verified by comparisons (figures 6-5, 6-6, and 6-7), but older cameos and suspected antique forgeries are often hopelessly confusing. Unless you are a scholar of Greek or Latin, the search for truth about signatures on antique cameos may prove to be an insurmountable challenge. Sometimes signatures simply cannot be verified, regardless of the depth of research on them. Signatures therefore, add value to the cameo only if they can be positively verified.

Two cultural factors on the value of modern cameos are fashionability and wearability. Fashionability takes into consideration the cycles in jewelry trends, caused by changing costume styles and the reactions of

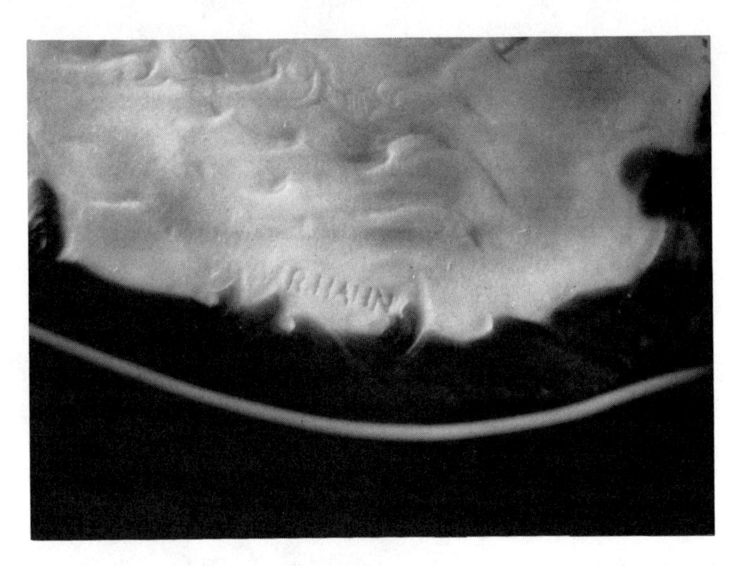


FIGURE 6-5.

*The signature of Richard H. Hahn.*

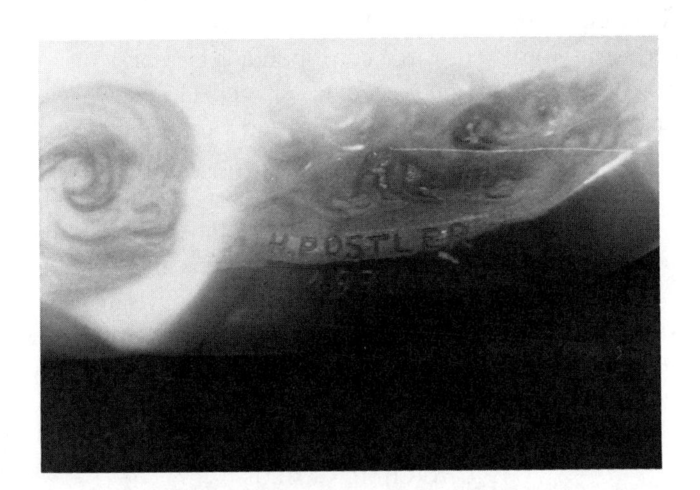


FIGURE 6-6.

*The signature of H. Postler. (From the collection of The Lizzadro Museum of Lapidary Art, Elmhurst, IL)*

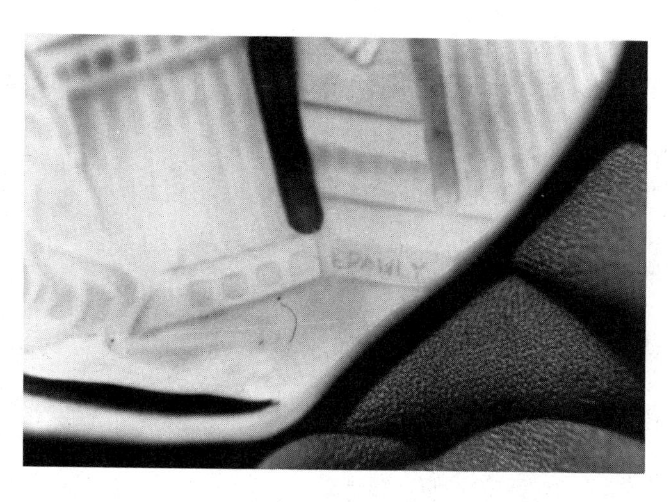

FIGURE 6-7.

*The signature of Erwin Pauly. (From the collection of The Lizzadro Museum of Lapidary Art, Elmhurst, IL)*

succeeding generations to the fashion tastes of earlier times. The vogue for collecting (especially antique cameos) is fueled in part by the costuming of the day. For example, the ladies' shawl, once popular, was an excellent garment upon which to fasten a cameo brooch. When the shawl declined in fashion, there was a corresponding lack of demand for the cameo brooch. Fashion trends seem to skip generations: while a second generation may reject some jewelry, third and fourth generations tend to look back nostalgically at the same pieces, and what had seemed ugly suddenly has appeal.

Wearability has to do with the nature of the materials that are in fashion. Obviously, some heavy hardstone cameos in oversized mountings cannot be comfortably worn on a flimsy, lightweight fabric. Shell cameos seem to fare better in the wearability category because they are of lightweight material and can generally be worn more comfortably.

## ARCHAIC CAMEOS

The established value parameters have already been discussed, but the collector should consider age and condition as separate elements when estimating the value of the most ancient cameos. Some appraisers may argue the point, but age alone is not the most important element in valuing an archaic cameo. It is, however, tied to its overall value because placement in time helps distinguish and identify it. Age *combined* with condition are the two principal forces that establish monetary value in archaic cameos. Other value elements may peripherally be applied, but their impact on value depends on the strength of the factor; an infallibly documented signature, for example, would definitely increase value.

*C h a p t e r*

# 7

# BUILDING A CAMEO COLLECTION

Why does anyone collect? The urge to collect is a primitive, deeply felt need to own objects for any of a number of reasons: usefulness, durability, rarity, social recognition, association with the memory of a person, place, or event, investment potential, and so forth. But, first and last, collecting should be *fun,* a pleasurable way to pass the time. With cameos there is an added bonus of being able to wear the collection pieces as ornamentation.

For most collectors the desire to build a serious collection of cameos does not rise full-blown like Athena leaping from Zeus's skull. The urge to collect, while instinctive in most humans—and some animals—creeps upon one gradually. First, a single cameo is purchased because it is interesting and has enduring qualities. Then another is bought because just looking at it, holding it, and studying it gives such joy. Another joins the group because its historic scene is as absorbing as a book. The cameo need not be ancient or antique, but it must have the capacity to enthrall and interest the viewer. In a short time, questions develop, causing the collector to expand his or her knowledge by studying the people, places, and events linked to the glyptic arts; the passion comes full circle as the cameo that provoked all the study becomes more fascinating than ever because of what has been learned about the glyptic arts.

The transition from bemused and detached spectator to anxious and eager collector takes only a short time and begins to gather momentum the very moment a friend or acquaintance politely inquires about one's interest in cameos.

One can specialize in collecting various periods, carvers, countries, or materials of cameos. It is not easy to collect archaic or ancient cameos, though, and membership in this dedicated collectors' group is very limited because of the availability—and cost—of these treasures. Even if one has the time, desire, and money to buy archaic cameos, many are simply no longer available for personal collections.

Once the rapture of purchasing the first few cameos subsides, the realities of cost and availability usually take hold. This applies even to those with unlimited means to pursue cameo collecting, because starting a collection makes demands—both financial and emotional—upon the individual. Once the mania for accumulating these treasures takes over, however, it is difficult to abandon, whatever the cost.

Many people believe that an antique is simply an item that is old. This is not the case, and collectors must grasp the terminology and understand the differences between archaic, antique, and old pieces. The U.S. Customs Bureau has established guidelines for antique items. To qualify as an antique (and therefore duty-free), the object must be at least one hundred years old. Anything from the far distant past, from ancient Rome or earlier, is considered archaic. Anything that is less than one hundred years of age is simply old; pieces denoted as modern or contemporary are post-World War II.

Collectors of cameo art are advised to concentrate on a particular theme, period of time, or carver. One of the oldest museum adages is to collect to strength, meaning that six pieces from a single master are worth more than six unrelated pieces. Careful planning marks the advanced collector, who sets boundaries on a collection and stays within them. Organizing a plan does more than just make the collector an expert; the collections themselves become more viewable and valuable and could well become a nucleus for future serious study by scholars. Also, one might consider a limit on the size of a collection, so that when the predetermined number is reached, the cameos of lesser value are sold. This upgrades and strengthens the group as a whole and keeps the collector active in the buying and selling market, in touch with other collectors, and up to date on prices and offerings.

For those with a desire to collect cameos and wise enough to distinguish collecting cycles, the 1990s is a choice time in which to begin building an outstanding collection. The more celebrated contemporary carvers in Germany and Italy are still living, and their works are relatively inexpensive, so *now* is the time to become a patron of their creative artistry. The names Richard H. Hahn, Manfred Wild, Erwin Pauly, Genaro Garofalo, Dante Sebastanelli, Salvatore Balzano, and many of their peers have been mentioned earlier in this book. A collection of the works of any one of them will almost certainly prove to be a worthwhile investment.

Maintaining interest, of course, is basic to the development of the collector, whether beginner or advanced. Scholars are quick to point out that a collector must also acquire and display the intangible attribute of taste. Glyptic scholars, such as Derek Content of Maine, point out that it is impossible to become a serious collector and make shrewd judgments without handling hundreds of cameos. This is not easy in view of the fact that, as Content claims, there are very few large and important private ancient cameo collections in the United States, with the exception of his own outstanding collection. So, a serious collector is hampered by the paucity of documented, historical, premium-quality cameos available for comparison. Collectors must depend primarily upon museum collections for study, even though they are not available for handling. Museum collections do allow comparison of periods, styles, motifs, and carving techniques because they are generally researched and labeled. However, they should be examined more than once: with each observation, a new and vital perspective is gained as the collector becomes experienced.

Unquestionably collecting is more fun when you know your subject. One source for viewing—and perhaps handling—antique and contemporary cameos is major public auctions, during the viewing exhibitions prior to sales. Never pass up an opportunity to see and examine different materials, even if the cameos turn out to be of limited importance.

A practical way to form a collection is through a knowledgeable cameo dealer. The strength of a dealer's expertise in this field is basic to a well-rounded collection. Those with knowledge are usually happy to share it with the serious beginner. It is safer to buy from known glyptic-arts dealers: they are jealous of their fine reputation as specialists and will be prepared to guarantee what they sell, backing it up with a certificate of authenticity.

If you find a cameo that you desire and are considering spending a large sum of money, do not hesitate to get the opinion of an expert *before* you buy. It is astounding how many people buy first, then seek an expert to confirm their "good deal." There is a caveat for the tourist or traveler looking to increase a collection while abroad: do not be tempted to pay high prices for an apparently genuine antique cameo just because you are vacationing in an exotic place such as Istanbul or Egypt. Even if a genuine article is offered, it is not necessarily for sale below the international market value. The world of antiques, especially glyptics, comprises a very small, close-knit group of buyers and sellers who know current market value all over the world. Insiders in buying ancient materials overseas advise that you are most vulnerable to being cheated when buying in the area where the antiques originate. Foreign dealers will freely guarantee that a cameo is ancient and authentic because they feel sure you are not likely to make an expensive return trip just to exercise the guarantee. A more urgent reason to refrain from buying any antiquities is because of the strict laws enacted by the Mediterranean countries against the export of their national treasures. Though it usually is not against the law for them to be offered to you for sale within their country of origin, they may not be taken out of the country.

When the collector begins to look for local cameos and starts to compare prices, he or she will discover they are more abundant in some parts of the country than others and that prices differ considerably. Regional preferences, as well as population size and economy, affect cost. New York, Boston, and Los Angeles are all cities where old cameos are relatively plentiful. As a result of the variations in demand, prices for old cameos may vary dramatically from one part of the country to the other, especially individual pieces offered by antique shops, flea markets, jewelry stores, and individual sellers. In general, dealers specializing in old or antique cameos know the market and have comparable pricing. Dealers of modern cameos have a common price structure, with variations caused by the individual dealer's overhead and markup.

A new collector can begin to understand more about cameo cost and pricing by compiling a price reference library. This is built with auction catalogs, dealer price lists, and a personal survey of cameos seen in antique shows. A simple spiral notepad can be kept for the express purpose of cost notation. Along with the price, it would be helpful to list the size and material of the cameo, subject, and condition. In this manner the collector has a guide to use when shopping for and comparing cameos. A *comparable* cameo is one of the same kind and quality of material, same size, same degree of carving skill, and approximately same age.

And finally, for collectors, a bit of sage advice from Gerhard Becker on the best way to build a collection and at the same time secure its value. "Always buy at the top," he suggests and adds, "If you have fifteen or twenty cameos, it may count as a collection, but if all are just average and lack any outstanding qualities, it may not be a very good collection." Becker counsels that quality is the key to a memorable collection. "It is better to have one really top-quality cameo than three secondary pieces."

Beginning collectors should review this checklist of questions before every purchase:

- Is it well carved? Complete? Damaged?
- Is it an original or a reproduction?
- Is it shell or stone?
- Is it ultrasonically carved?

- Does it have a signature?
- Does the subject have a nice visual impact?
- Is it a good example of a particular subject, artist, era, or material?
- Is it better than average?

## BUYING AND SELLING CAMEOS

When buying either a hardstone or shell cameo, one of the most important tools to own and use is a 10X jeweler's loupe (pronounced *loop*) to examine the details of a cameo and inspect the surface. One of the first items of business when buying a cameo is to run a fingernail around the edge to feel for cracks and undercutting. This exercise needs to be combined with a loupe for a complete inspection. A loupe can also help in distinguishing between ultrasonically carved and hand-carved cameos and to spot indications of forgery. All buyers should learn to use a loupe like an expert. The most convenient type to use is a triplet lens that has been corrected for image distortion and color fringing; these are available at most jewelers' supply companies. Using a loupe is not difficult, but it may take some practice to understand exactly what is being seen. These are the basic steps in using a hand loupe:

1. Hold the loupe between the thumb and forefinger, with the viewing lens facing away from your body.

2. Hold the cameo in your other hand.

3. Move the hand holding the loupe as close to your eye as possible. If you wear glasses, you do not have to remove them, just rest the loupe against the lens.

4. To steady the hand holding the loupe, brace your arm against your chest.

5. Move the cameo slowly toward the loupe until it is in sharp focus, usually 1 inch or less from the loupe.

### BUYING AT AUCTION

Auction sales can be one of the most fruitful means of finding cameos to build a collection. The major auction houses, such as Christie's and Sotheby's, offer catalogs of sales on a subscription basis or for specific sales. But they are not the only companies handling jewelry auctions that include cameos. At dozens of regional galleries all over the United States, cameo jewelry changes hands, generally at prices that satisfy all parties (see the appendix for a list of galleries).

The collector who buys at auction must be aware that every auctioneer may not be as knowledgeable as he or she should be about glyptics, and one auction may not be as well attended as another, better-advertised one. These factors can mean either a good deal or mistake for the buyer. Those familiar with this territory can find bargains, but it may be necessary to attend many auctions and be fully conversant with the stock and the styles, attitudes, and knowledge of individual auctioneers before getting the prize. Consider, for example, the subject matter used on cameos. The ardent collector will know much more about myths and portraits than the average auctioneer or the average appraiser for the auction house will. While auction houses do try to cover all bases to present the merchandise in the most favorable light—to get the best price—sometimes they slip up, and when they do, it is usually in the area of design, or subject. Mostly, they just do not know what an image represents.

Those familiar with Greco-Roman mythology know that Athena (Minerva) frequently appears with a bird beside her, an owl; Venus is usually shown with her baby Cupid; and so forth. As an amusing challenge, browse through auction catalogs looking at photos of cameos that have been or are currently for sale. To your astonishment, you will find that you often know more about the subject than the person who captioned the picture. In many cases cameos are misinterpreted. For example, a major auction house recently showed a gold and stone cameo brooch with a mid-nineteenth-century label and the following description: "Depicting an allegorical scene of two maidens overlooking a putto floating in the clouds, holding aloft a flower-filled cornucopia, as an owl surveys; the reverse signed F. Zignani, Engraver, Roma."

The cameo was sardonyx and set in an 18-karat gold mounting as a brooch. The piece sold for an amazing $577; it was worth much more. And it might have sold for more if it had been more precisely presented. The circa dating was accurate, but there was no mention in the description that the piece was in mint condition and masterfully carved, important factors that might have tempted more bidders. The scene itself was misconstrued; far from being simply allegorical, it consisted of portraits of Venus and Cupid with Minerva and her owl. Sky symbols, the evening star and quarter-moon, also provide subject identification, a point that was apparently missed when the picture of the cameo was captioned. Auction catalogs are filled with such omissions. The opportunities for the collector to obtain an excellent cameo at a small price are better than ever.

On the other hand, cameos with well-researched provenance that are signed by known carvers are given good publicity by an auction house receive high bids. This benefits the collector as well, because it provides a chance to see some of the best cameos available while providing sales validation of prices that the collector can use later as comparables. Because most cameos to be auctioned are mounted in jewelry, it must be kept in mind that some will bring a big price because of their mountings, not because of the cameos themselves.

The name Castellani is magic at auction sales. Fortunato Pio Castellani (1794 to 1865) was founder of a Roman jewelry shop that boasted such celebrated clients as Napoléon III of France, Prince Albert of Great Britain, Empress Victoria Adelaide of Prussia, Queen Victoria's daughter; King Victor Emmanuel II; and Robert and Elizabeth Barrett Browning. The company included his sons Augusto and Alessandro and was operated by his grandson Alfredo until his death in 1930. Castellani had branches in Paris and London. The firm was widely known for copies of ancient Greek and Roman designs and for the use of genuine antique cameos and gems incorporated into their own mountings. In a 1989 sale at Dunning's Auction in Elgin, Illinois, a hardstone white and tan cameo with a portrait of Medusa, 33 millimeters, brought nearly $12,000 (fig. 7-1). The generous price was mostly attributable to the yellow-gold archaeological style of nineteenth-century revival brooch mounting hallmarked with the back-to-back intertwined *C*s of the Castellani workshop. Making it even more attractive, the brooch came complete with its original fitted velvet box.

FIGURE 7-1.

*This Castellani brooch with a hardstone cameo portrait of Medusa brought close to $12,000 at auction. (Courtesy of Dunning's Auction Service)*

A few years earlier, a carved moonstone cameo of Queen Elizabeth I set in a diamond and demantoid garnet mounting brought over $13,000 in a California auction. What the two auctions had in common was an excellent catalog presentation that drew a crowd of interested buyers.

The collector's market for cameos, both loose and mounted in jewelry, is active and growing stronger. If you want to be a player in the auction game, now is the time to study, research, and scrutinize the cameos at every auction, small and large.

There are caveats when buying at auction. Realize that some auctioneers are willing to mislead you into believing that something is what it is not. For instance, if you are looking at so-called antique cameos, insist on close examination with a loupe, and do not forget to ask pertinent questions about the mountings if they are set; for example, ask what the metal is, what the karat content is, and what the piece weighs in grams. Remember the definition of antique, and inquire if the piece is over one hundred years old. If the answer is yes, ask if they will give you a receipt identifying it as such if you are the lucky bidder. Beware of inaccurate or misleading descriptions of cameo materials, such as "German lapis," a term that is used for dyed blue jasper, or "Manchurian jade," which is actually the mineral soapstone.

If you are compiling your own price guide with auction sales prices, make notes about the conditions surrounding the sale because location, time of year, weather, holidays, and publicity all affect final prices. A poorly attended auction, resulting in lower-than-expected prices, should especially be noted because the prices may not accurately reflect the average market value. You will then be prepared to pay more, or less, for a particular cameo, depending upon the external conditions of the next auction you attend.

---

## Selling at Auction

If you want to sell a few items or your entire cameo collection, you must invest as much energy into its disposition as you did in its acquisition. And you must be prepared to be absolutely realistic in your pricing. The basis for pricing cameos put up for auction must be the collector's research of the marketplace, not fantasy or dreams of what they might bring.

Whether sending one piece or an entire collection to auction, insist upon a reserve price, the lowest price you will accept for the piece; if the bidding does not reach the price, the item is returned to you, and you pay a commission. The reserve price is established by you and the auction representative *before* you sign a contract of consignment. Auctioneers normally charge both buyers and sellers a commission, about 10 to 15 percent of the realized sales price. Other fees that must be paid by the seller include insurance to protect merchandise while it is in the auction company's hands and a photographic fee if the piece is to be displayed in a catalog.

Before contacting the auction representative, make a list of the qualities that make your cameos valuable. Do not forget about timing and trends, such as fashion and nostalgia for certain colors and gemstone materials. The factors to consider are the same as those discussed earlier: age, beauty, collectibility, materials, signature, provenance, period, style, rarity, workmanship, and so forth. Note any significant strength of the cameo or collection and describe how it adds to the desirability, so it can be cataloged to best advantage. Note if any cameo has been associated with a well-known personality or event. Then, consider how all these factors might attract a specific type of buyer, and describe that buyer—perhaps an individual collector or a museum. Communicate all this information to the auction representative. An entire collection brings more money than individual items from the collection sold singly do.

## SELLING ON YOUR OWN

If you prefer to try your own luck at selling your collection, make a list of potential customers: antique dealers, individual collectors, used-jewelry retailers. Then, record the elements of value as you would when selling at auction. Use the information to test the reaction of collectors or dealers to buying this type of goods. If you know the price you want for the cameos, make a list of the pieces and prices and send it around to other collectors, retail jewelers, and antique dealers who specialize in cameos. Then follow up with a telephone call or personal visit to those who have expressed interest. Collections have changed hands quickly in this way; all it takes is one interested buyer and one serious seller.

## PROVENANCE

Provenance is more than just a fable about a stone's background; it is documented evidence of the history and origins of a piece. In fact, a glyptic-arts expert said this about how he avoided purchasing forgeries: "I never buy stories." He is not taken in by the hype or romance often attached to a gem or piece of jewelry for sale; he judges with clear, cool detachment. Unless the history or origin of the piece has been fully documented and can be substantiated, it is discounted.

This advice applies not only to the many stories spun about an item's past but also extends to the source of the supply. Reliable sources can be mistaken; offhand speculation that a cameo is from Napoléon's extensive collection does not make it genuine. Interesting narratives of provenance should not affect one's final decision about purchasing a cameo, even if they are told in luxurious surroundings. Handsome offices or imposing shops do not necessarily give the cameo an illustrious past, do not guarantee that its origin has been verified or that the salesperson knows what the gem is about. If the origin of the cameo cannot be documented in writing, discount it as gossip.

Auction companies capitalize on sound provenance by using it in their catalogs, because they know it increases value, especially of antique jewelry. The lovely gold and hardstone cameo comb from early-nineteenth-century France shown in figure 7-2 was listed in an auction catalog a few years ago with the following provenance:

*Four-color gold, decorated with cannetille work and set with 7 oval hard-stone cameos depicting various portraits, the back engraved "Pauline*

FIGURE 7-2.

*The Pauline Bonaparte Borghese cameo comb has well-documented provenance. (Photo © 1981 Sotheby's, Inc.)*

*Bonaparte Borghese" between the Borghese eagle crest and the Imperial coronet, with yellow metal tines, has later leather box with original fitted form covered in velvet.*

*Pauline Bonaparte, Napoléon Bonaparte's sister, was first married to General Charles Le Clerc, who died six years after their marriage. Her second marriage was to the Italian nobleman Prince Camillo Borghese.*

This expert description and skillfully revealed provenance produced hot bidding. Some names are magic; Bonaparte is one.

---

## Appraisals

If you have a growing collection of cameos, you may at some time need an appraisal. Any large collection should be insured against theft, loss, or damage, and the insurance company will require an estimate of the current replacement value. Obtain an "Agreed-Value" policy, which guarantees that in case of catastrophe you receive the full monetary value stated on your scheduled items, not a discount of that figure. Obtaining a comprehensive appraisal can be an expensive undertaking, but it is highly recommended if you have numerous, antique, or rare cameos.

An appraisal of a glyptics collection cannot be satisfactorily handled by a local jeweler or antique dealer. It is too specialized and requires a qualified expert in the appraisal field—there is no such thing as a Blue Book value for a cameo collection. Estimating value demands knowledge and research.

The need for an appraisal may also arise when one wants to buy or sell an especially fine cameo. The buyer wants a document to warrant the cameo as genuine (age or material); the seller wants a report to show a prospective buyer in order to get the best price for the gem.

Appraisal certificates are not as necessary when buying modern or commercial cameo jewelry, unless the cameo is from the hand of a noted contemporary carver. They are for archaic, antique, or the very expensive rare cameos. It is not unusual for a Greco-Roman cameo in good condition to sell for $20,000 or more. When a cameo attains this value, it seems sensible to get insurance to protect it as well as a certificate to authenticate it.

It should be noted that there is a difference between an insurance appraisal and a certificate of authenticity. A certificate of authenticity provides a complete description of the cameo with measurements, weight (or estimated weight if it is mounted as a piece of jewelry), identification and quality grade of the material, circa date, country of origin (if it can be determined), motif or design, provenance (if available), evaluation of craftsmanship, validation of signature (if signed) or an opinion as to the item's style or school of art. A certificate provides much information but usually includes no market price or estimated value. On the other hand, an insurance appraisal is an evaluation of the cameo, including facts about its purchase or acquisition, if available, and an estimated cost of its replacement, to protect the owner's investment if the cameo is lost, stolen, or suffers damage. A well-researched appraisal may be expensive, but it will be worth the cost, because it documents the history of the cameo, its complete description, and the current market price.

## CAMEOS AS INVESTMENT GEMS

The market for investing in jewelry, gemstones, and objects of art has grown tremendously since the 1970s, fueled in part by the potentially high returns such investments may bring. Some economists have even described fine jewelry (including engraved gems) as a "hard asset," with the value increasing about 15 percent per year.

This is a dangerously speculative pronouncement, made by people who do not understand the complexities of the gem collectibles market and attempt to treat gems like gold, silver, and platinum. The gem market is entirely different from the precious-metal market. Though gems may retain or increase their value as long as the standards upon which it is based remain solid, values can change with lightning speed, often overnight. These assets cannot be immediately liquidated, and a sudden decline in their value will therefore likely spell disaster for the "investor." Changes in the value of gemstones and engraved gems—for better or worse—may have no obvious relationship to any intrinsic value standards and can occur at any time.

Cameos and seal stones, like other precious gems, should never be bought strictly as investments. Even though they *might* be a hedge against inflation, when purchased through the normal retail-market channels, the markups are substantial. The gem passes through many hands from carver to marketplace, and a collector will have to wait a long time, perhaps a decade or more, to recapture the original purchase price.

One way to estimate whether or not your cameo purchase has

| Year of Purchase | Inflation Factor | Year of Purchase | Inflation Factor |
|---|---|---|---|
| *1967* | *4.07* | *1979* | *1.70* |
| *1968* | *3.80* | *1980* | *1.55* |
| *1969* | *3.55* | *1981* | *1.41* |
| *1970* | *3.32* | *1982* | *1.29* |
| *1971* | *3.10* | *1983* | *1.24* |
| *1972* | *2.89* | *1984* | *1.20* |
| *1973* | *2.70* | *1985* | *1.16* |
| *1974* | *2.52* | *1986* | *1.12* |
| *1975* | *2.35* | *1987* | *1.08* |
| *1976* | *2.20* | *1988* | *1.05* |
| *1977* | *2.05* | *1989* | *1.00* |
| *1978* | *1.87* | | |

*Source:* Compiled from figures from the Bureau of Labor Statistics.

matched or surpassed the rate of inflation (and grown in value) is to take the original dollar amount that you paid and adjust it for inflation to the current value of the dollar. Use the information in the following table; with some basic math, you can calculate inflation-corrected prices for your cameos. To determine how much your cameo or entire collection should bring today, find the year that you bought it. Multiply the amount you paid for it by the inflation factor for that year. For example, a cameo purchased in 1967 for $100 would now have to bring $407.00 (4.07 times $100) to have kept pace with inflation. With the figures to guide you, you will be able to determine the ***minimum*** price to set if you wish to sell, though you may want to ask a higher price.

## Care and Conservation of Cameos

Cameos, if not worn as jewelry, should be displayed in trays or drawers with many lined compartments. Coin cabinets are ideal display units for smaller cameos. Although lighted cabinets will not affect hardstone cameos, they may damage collections of other materials. Delicate shell, coral, and ivory will dry out and become brittle when exposed for any length of time to the heat of a lighted cabinet. If you hold a shell cameo up to a light and see faint vertical lines in the translucent material, they may be the beginnings of cracks that cannot be prevented. Therefore, take no chances with shell cameos; keep them away from heat, and provide enough room in the display so they are not crowded. Hardstones that are translucent (such as some amethysts) look more beautiful when they are lit from the back, as the color of the material is shown to greatest advantage.

All cameos should have explanatory labels. The serious collector keeps a card file of the collection, noting the dates and places of acquisitions and prices paid.

---

## Cleaning

Hardstone is not immune from environmental attack by fungus, mildew, and other surface-layer contaminants, including smoke, fumes, dirt, and household dust. But cleaning must be approached with great caution. The cameo must never be subjected to harsh methods.

There is hardly a collector who has not at some time looked at a cameo and wondered how to clean it—and some have tried, with tragic results. The best advice may be simply to do nothing, or to take the cameo to a specialist (ask a museum to recommend a reputable specialist). Ultrasonic cleaning and clumsy home cleaning should not be attempted. However, since most collectors want to keep their treasures in the best possible condition, the following suggestions may prove helpful.

First, the cameo should be thoroughly dusted with a soft brush, such as an artist's sable brush, paying attention to the details of the modeling, where dust often accumulates. If the dust is left on the material, it will work deeper into the detail and scratch the surface; dust frequently has a hardness greater than that of the cameo material. After dusting, rinse the cameo, using a minimum of liquid. Some specific tips follow.

*Hardstones* require various treatments. Only the hardest stones, such as diamond, ruby, and sapphire, can be cleaned with an ultrasonic cleaning machine or a commercial jewelry cleaner. All other stone cameos should be kept away from any harsh cleaning treatment, spirits, solvents, or other chemicals; just dust, rinse with tepid water, and gently dry. Porous stones, such as turquoise, are especially affected by moisture and suffer permanent color change if they absorb oil or, often, even water, because water is laced with chemicals. The once beautiful blue-green turquoise will slowly turn a dingy gray green. Turquoise cameos should *only* be dusted lightly with a soft brush.

*Shell* is damaged easily and should never be cleaned with anything other than tepid water. Dust, rinse several times, and carefully pat dry with a cotton cloth. Shell is more easily scratched by dust and dirt than stone is.

*Mother-of-pearl* carved into a cameo is cleaned with a cotton cloth dipped in a mixture of mild soap suds and tepid water. The piece should then be carefully patted dry with a soft cotton cloth.

*Amber* cameos are popular. Because the material is very soft and has electrostatic qualities, dust and dirt particles accumulate in the details of the modeling. Never clean amber with any type of solvent, even something as mild as a turpentine substitute; amber is a fossilized tree resin, which will dissolve when treated in this manner. Gentle, repeated rinsings in tepid water followed by gentle patting with a cotton cloth to dry are suggested.

*Plaster casts,* as well as bas-relief carvings in natural materials, require care when cleaning. This friable material is very delicate and highly absorbent. *Never use liquid.* To clean, the cast must be totally dry. It is brushed with an artist's soft brush that has been dipped in dry plaster of paris.

*Ivory* behaves like wood when it comes into contact with moisture and heat: the material can separate and crack if it swells unevenly or distort and warp if it becomes too hot. With ivory, the greatest care must be taken after dusting and a quick dip in a warm water bath; the piece must be well dried immediately with soft cotton cloth. Ivory often develops a yellowish tinge that increases with age and cannot be reversed. This is a natural patination, the result of time, not a defect.

*Glass* cameos also deteriorate. Experts that restore old glass recommend keeping it away from dampness, air conditioning, and heating vents. Storage cabinets in which old glass cameos are kept should also be well ventilated. Brushing is the only recommended cleaning method.

---

## Cameo Damage and Repair

It is not easy to fight the effects of time, heat, cold, careless handling, and abuse on individual cameos or entire collections. Cameos do fall victim to scratches, chipping, and corrosive chemicals. In some of the older pieces, especially shell cameos, the subject matter has become completely indistinguishable because of improper handling or frequent wear.

A cameo is more likely to be damaged if it is undercut. When a cameo is deeply undercut and in high relief, there is always a danger that part of the subject will break away, especially raised parts, such as arms, noses, swords, and helmets. When breakage occurs, the work loses value because a broken stone cameo cannot be effectively repaired. If it is a newly carved piece, at least half of its value is gone. A broken background, or the broken margin of a cameo, can be subjected to some slight cosmetic repair, using a cut-to-fit stone to fill in the lost portion of background or subject, but the repair will always be noticeable. Some cameos, especially those with intrinsic value, have the lost portion of the background filled in or covered over with gold when placed in a mounting. This helps mask the defect to a slight degree.

Because antique cameos frequently have historical value, they should be salvaged by any possible means. An archaic cameo is so historically valuable that, regardless of chips, breaks, or cracks, it should be carefully and conscientiously preserved at all costs (fig. 7-3).

Broken shell cameos cannot be repaired, but they might be able to take on a new form, if the break is in the background, as in the cameo pictured at the bottom center of figure 7-4. A cameo portrait can be cut out and away from its background by a skilled lapidary and then converted into a free bas-relief form. It can then be fitted with a pendant or brooch mounting, like the portrait at the top center of figure 7-4.

FIGURE 7-3.

*A broken amber cameo picturing a sleeping nymph and a satyr stealing up on her, with Cupid aiming an arrow at the satyr. The cameo has been dated as sixteenth century and is important enough to be conserved. It has been placed in a frame to hold the pieces together until they can be professionally mended. (The Sommerville collection, University of Pennsylvania)*

FIGURE 7-4.

*Broken shell cameos (center bottom) are impossible to mend, but transformation is a solution. The cameos can be cut from the background and fitted with a mounting (top center) that will give the piece a new life.*

How much you can do to repair a broken cameo yourself depends upon the material and the value of the piece, both on the market and to you personally. A fine antique should be mended by a professional, but if the cameo is not valuable enough to justify the expense of a conservator, yet too treasured to be discarded, you may want to attempt some kind of repair. If so, use as simple a material as you can and some type of cement or fast-acting glue for repairs. Spend time to produce the smallest, smoothest glue joint possible; while it will never be invisible, with care the piece may be presentable.

Chapter

# 8

# GREAT AND HISTORIC COLLECTIONS

Collectors of archaic, antique, or modern cameos who wish advice and information from museum curators on building or upgrading their collections should not be discouraged upon seeing the extraordinary quality of the specimens in some exhibits. Museum collections have grown over the generations with the investment of enormous sums of public, and often private, funds; many have grown from the acquisition of private collections, the majority of which were never for sale on the open market and therefore could not have been purchased at any price.

The most important public collections of cameos and engraved gems are administered by staff trained in geology, gemology, or art. All museum personnel contacted in the research for this book were willing to assist collectors in every possible way to increase their knowledge. Because only a very small percentage of the total collection of cameos and other engraved gems of an institution is on display at any one time, a collector who wishes to examine the gems remaining behind the scenes will find it necessary to make application and receive permission in advance before any examination can take place. One cannot just walk unexpected into a museum and ask to examine a valuable collection. Specific access must be granted, sometimes by more than one department head. Also, someone on staff must be on hand to serve as your escort during the visit.

The card catalog department can also be accessed to review information and photos of existing collections. Admission to the card catalog department also requires advance permission.

Some important public collections of cameos, seals, and intaglios in the United States are found in the following museums: Boston Museum of Fine Arts, Boston; J. Paul Getty Museum, Malibu, California; Indiana University Museum, Bloomington, Indiana; Lizzadro Museum of Lapidary Arts, Elmhurst, Illinois; Metropolitan Museum of Art, New York; University Museum, University of Pennsylvania, Philadelphia; and Walters Art Gallery, Baltimore, Maryland.

Distinguished collections can also be found in these foreign museums: Antiquities Museum, Cairo; Archaeological Museum, Istanbul; Cameo Museum, Yamanashi-ken, Japan; Ashmolean Museum, Oxford, England; British Museum, London; Bibliothèque Nationale, Paris; the Louvre, Paris; Deutsches Edelsteinmuseum, Idar-Oberstein, Germany; Hermitage, Leningrad; Kunsthistorische Museum, Vienna; Museo Nazionale, Florence; Royal Library, The Hague; Staatliche Munzsammlung Museum, Munich; and Thorvaldsen Museum, Copenhagen.

The Boston Museum of Fine Arts has a small but well-crafted collection of cameos of the eighteenth and nineteenth centuries. Most are portraits in hardstone from the early and mid-nineteenth century. One fine gold bracelet set with four cameos carved by William Morris Hunt (1824 to 1879), circa 1840, that portrays the four Hunt brothers is noteworthy.

The J. Paul Getty Museum is located on the Pacific Coast Highway in Malibu, about twenty-five miles west of downtown Los Angeles. It houses permanent collections of Greek and Roman artifacts, pre-twentieth-century European paintings, drawings, sculpture, illuminated manuscripts, decorative arts, and of course, a small but excellent collection of cameos, intaglios, and cameo glass. The building itself is a re-creation of a first century A.D. Roman country villa, complete with interior and exterior gardens.

The museum has more than 450 engraved stones, most with mythological scenes, but also including gems with magic inscriptions engraved during the Gnostic period.

One that elicits acclaim from all viewers is a carnelian intaglio attributed to the Greek carver Epimenes, 500 B.C. The piece measures about ⅓ by ¼ by ½ inch (8 by 6 by 11 millimeters) and has a scarab engraved on the domed top. On the reverse is a carved image of a youth leaning on a stick as he bends over to fasten the strap of his sandal. The piece has been praised for its delicate treatment of the hair and facial features and well-defined musculature.

Another notable piece is a garnet bust of a Hellenistic queen, about ¾ by ½ by ¼ inch (19 by 13.8 by 6.3 millimeters), labeled as Greek from the third century B.C. The museum description says the shallow relief probably depicts the wife of the Egyptian king Ptolemy III, Berenice II, which is substantiated with the information that the portrait is similar to those on coins issued during the reign of her husband, from 246 to 221 B.C.

The agate cameo of the Emperor Aurelian is Roman, circa A.D. 260 to 280. This is a very small cameo, ¾ by ½ inch, with high-quality material and outstanding workmanship. The style of the craftsmanship suggests it may have been engraved in an imperial workshop, as the facial features are smooth and carefully modeled, with a short beard and close-fitting cap of hair. The museum journal points out that the counterpoint is emphasized by the engraving technique, in which long strokes are used for flesh surfaces and short strokes for hair and beard. These, the writer believed, were wheel-generated.

The Indiana University Museum in Bloomington is host to the esteemed Burton Y. Berry Collection of ancient gems, on extended loan to the facility. Berry collected classical antiquities during his employment with the U.S. Foreign Service while on assignment to the Balkans and Near East. The ancient gems on view are some of the exquisite and rare intaglios and cameos he assembled in carnelian, jasper, emerald, agate, garnet, and other hardstone materials. They range in period from 1700 B.C. to A.D. 1800.

The Roman cameos are especially interesting and worthy of study for their style and modeling. An oval sardonyx, about ¾ by 1 inch (19 by 27 millimeters), depicts Athena standing, writing on a tablet dedicating a panoply hung on a tree trunk. There is also a two-layer onyx depicting Nike as Eos, driving a two-wheeled chariot, and a three-layer onyx bust of Psyche with short curly hair bunched on her head, held in place with a fillet. Also in the collection are: a four-layer sardonyx cameo with the head of a Gorgon; a two-layer onyx cameo portrait of a man in profile with a beard, wearing a wreath around his head; a five-layer onyx cameo, about ½ by ½ inch (14 by 15 millimeters), of the conjugated heads of two men. The collection also includes several exceptional agate motto cameos.

The Lizzadro Museum of Lapidary Art is near Chicago. This public museum was founded to house and display gems, minerals, and art objects made of gem materials and to promote interest in the lapidary arts and the study and collection of minerals. It is unique in the United States. The Lizzadro is home to a world-class collection of jade objects of art and jade carvings, but it has a prominent cameo collection as well. The Lizzadro's cameos number into the hundreds, in shell, agate, amethyst, onyx, malachite, tigereye, and some rare materials. Many are mounted in gold and set with pearls as pieces of jewelry. One malachite chevee—a cameo carved in the center of a bowllike depression—is set in gold and surrounded by forty-four pearls. Along with portrait cameos, the museum has a fine collection of cameo agate plates and bowls, most with subjects combining symbolism and myth, all modern works from Idar-Obserstein (fig. 8-1). New pieces are added from time to time.

FIGURE 8-1.

*This fine agate bowl is a recent acquisition of the Lizzadro Museum. The work, 10 inches in diameter and 2 inches deep, was executed in 1970 by H. Postler, who signed the vessel. The subject is Poseidon and Amphitrite, with Triton blowing his horn. (From the collection of The Lizzadro Museum of Lapidary Art, Elmhurst, IL)*

The Lizzadro was fortunate to acquire the collection of the late Dr. J. Daniel Willems, a Chicago physician and zealous cameo collector. One particularly conspicuous shell cameo from his cabinet is the portrait of an Italian lady that Dr. Willems liked to call his "Italian countess" (color plate 60). Actually, the cameo is a copy of a painting by Antonio Pollaiuolo that hangs in the Museo Poldi Pezzoli in Milan, of the wife of Giovanni Bardi, an Italian nobleman, but Dr. Willems did not know this. He was so enthralled with his "countess" that he discussed her at length in his numerous writings on cameos. In one article he wrote, "The carving is exquisite with the beautiful young woman in lifelike colors in classical profile and dressed in regal finery. The outstanding feature is her slender neck with its chain and pendant. The links in the chain are each carved in equal size, the whole chain painted in such fine detail that each link can be counted individually." Willems collected for scores of years, but the cameo he called the "countess" was his prize. Of her artistic qualities, he said, "The cameo is a marvel without a flaw. The colors are soft and warm, the profile is perfection. How did the artist do it? Did he know the lady personally? He must have cut it under a magnifying lens, using for the painting an eyelash hair, which is rigid and tapers to a fine point. Carefully dipped in a fluid color, no drop would fall except the tiniest dot needed precisely on the right spot on the cameo."

Willems also liked multilayered agates, such as those in figures 8-2 and 8-3, with classical subjects, as well as trendy cameos, such as the naturally colored black, pink, and white shell cameo carved with a portrait of a turbaned Nubian (fig. 8-4). The black layer contains the face of the Nubian in profile, carved masterfully and the other colored layers are put to good use as the turban and jewelry ornamentation. The cameo is mounted in a setting as a locket, with a deep compartment like the type that was used to hold the hair of a loved one. The back of the mounting is engraved: "Mother from Newton, 1868."

The museum's collection of shell cameos includes a complete set of the fourteen stations of the cross, illustrating Christ's tribulations before his crucifixion. Each of the cameos is an oval, about 1½ by 2 inches (40 by 50 millimeters). An even larger oval depicts the Last Supper, after the Leonardo da Vinci painting (fig. 8-5). Religious-themed cameos are considered collectibles by the Lizzadro, and the museum cabinets contain several examples of the portraits of saints and other religious subjects. All of the shell cameos are contemporary works.

The Metropolitan Museum of Art in New York is home to several outstanding collections of cameos, intaglios, and seal stones. One especially impressive collection devoted to postclassical cameos is the Milton Weil Collection. Because the exhibits rotate, many individual pieces from this particular acquisition may not be on display. If examination is desired by the scholar or student, a letter of application to view them is necessary.

The Weil collection is important in that it shows the interpretation of the classical tradition in more modern times. Late Roman, Byzantine, and medieval periods are represented but by only a few examples; the majority of the cameos are from the Renaissance.

One of the most interesting works in the Weil group is a double cameo portrait executed by sculptor Leone Leoni as an experiment. It was his first and only cameo engraving. Leoni himself described the project as a caprice, to which he expected to devote no more than three months of his time. One side of the sardonyx cameo contains the portraits of the Emperor Charles V and his son King Philip II; the thunderbolt, associated with Zeus, is behind the head of Charles. On the reverse is a portrait of the Empress Isabella, probably after a painting by Titian. A bronze cast of the original is in the Munzsammlung in Munich.

Many artists of the eighteenth and nineteenth centuries are repre-

FIGURE 8-2.

*This finely crafted three-layer agate cameo picturing Apollo with wings is on display at the Lizzadro Museum, in the Willems collection. (From the collection of The Lizzadro Museum of Lapidary Art, Elmhurst, IL)*

FIGURE 8-3.

*A depiction of Flora in a three-layered agate, from the Willems collection at the Lizzadro Museum. (From the collection of The Lizzadro Museum of Lapidary Art, Elmhurst, IL)*

FIGURE 8-4.

*A remarkable cameo of a Nubian in natural black, white, and pink shell, in the Lizzadro Museum. (From the collection of The Lizzadro Museum of Lapidary Art, Elmhurst, IL)*

FIGURE 8-5.

*A shell rendition of Leonardo's* Last Supper, *at the Lizzadro Museum. (From the collection of The Lizzadro Museum of Lapidary Art, Elmhurst, IL)*

sented in the Weil collection, including signed cameos by Amastini, Girometti, Natter, Pichler, Pistrucci, Rega, and Saulini.

The museum at the University of Pennsylvania in Philadelphia has, among other collections, the cameos and intaglios of Maxwell Sommerville. Sommerville was professor of glyptology at the university, and it was his bequest that swelled the museum's cameo collection to one of the largest of ancient and later engraved gems in the United States. It is presently not on public view.

The Sommerville gems, over three thousand items, reflect the tastes of one nineteenth-century connoisseur who worked diligently to gather a grand cabinet in the tradition of the nobility and aristocracy.

FIGURE 8-6.

*Two neoclassical cameos in the Sommerville collection.* Top: *The vestal virgins guard the Palladium temple.* Bottom: *Marcus Aurelius addresses Roman officials and soldiers. The scene is based on a relief from a lost triumphal arch. The cameo has a circa date of 1700 to 1850. (The Sommerville collection, University of Pennsylvania)*

Sommerville assembled Babylonian cylinder seals along with Persian seals, Greek stones, Etruscan scarabs of the sixth to fourth centuries B.C., and cut stones and ancient pastes of Greece and Rome from the fourth to the first century B.C. His eclectic collection also contains many stones of later periods, from the medieval Byzantine and Western European to neoclassic Italian.

Sommerville wrote two books and numerous magazine articles extrolling the virtues of his gems. He liked to point out the neoclassical cameos with subjects based on historic events (figs. 8-6 and 8-7). As a deeply spiritual man interested in religions of the world, he enjoyed musing about mythological subjects, legends, and fables (figs. 8-8, 8-9, and 8-10). Portrait collecting especially seemed to fascinate him (figs. 8-11, 8-12, and 8-13).

Sommerville referred to his cameos as "pages in my stone books" and patiently tried to read and record the message conveyed by each cameo. Believing his cameos to be transmitters of ideas and recorders of emotions, he wrote, "Almost no plea has been offered for glyptology as a factor contributing (to) historical data. The mass of scientists have been contented with musty old volumes, and these little message-bearing stones have been regarded as nothing more than curious ancient articles of luxury, yet we recognize each and every piece as part of a great story, recording and illustrating many epochs and eras in this world's history." The final passage in one of Sommerville's writings was this comment: "We have been seeking to replace each fragment into its proper place . . . we

FIGURE 8-7

*Chalcedony cameos based on the triumphs of Trajan, from the Sommerville collection. (The Sommerville collection, University of Pennsylvania)*

**FIGURE 8-8.**

*Professor Sommerville enjoyed collecting cameos with mythological subjects. This neoclassical agate cameo has a Greek inscription: "Of the Sweeter Deities." (The Sommerville collection, University of Pennsylvania)*

**FIGURE 8-9.**

*A sardonyx cameo of Meleager and Atalanta dancing. (The Sommerville collection, University of Pennsylvania)*

---

are convinced that we read thereon many things that no manuscripts or books have communicated to us."

As a consummate collector, Professor Sommerville had no equal, and he was as colorful a character as any gem in his cabinet. Following his death, the collection he had so carefully assembled for decades became the object of attack. Many of the pieces were said to be forgeries and were

FIGURE 8-10.

*This neoclassical chalcedony cameo shows Venus, Jupiter, and an attribute of Jupiter, the eagle. (The Sommerville collection, University of Pennsylvania)*

carefully scrutinized by the board of managers of the museum. The questions about some of the gems were made public by S. Hudson Chapman, a numismatist, after some investigation of authenticity by experts and scholars. Although the full outcome of their findings remains muddled, there is no doubt that the collections are most certainly antique gems in this century, of immense value for educational studies.

Sommerville was a curious and trusting scholar. A slightly built man of frail appearance with a receding forehead, he sported round wire spectacles and a neatly groomed Vandyke beard and mustache. Sommerville delighted in dressing up in his authentic Buddhist robes and ornaments and was often photographed wearing them posed against a background of his curios. In an article dated December 29, 1905, in the Pennsylvania newspaper *The North American,* he was described as having a "simplicity of character and trustfulness of nature." The assessment of his character was in defense of the accusations about the supposed worthlessness of his collection. The professor's own words were quoted, explaining how he had collected talismans, engraved and carved gems, during one of his longest and most productive foreign buying expeditions:

*I hardly know how many or what objects I have acquired at this time. They are different from most of the other gems in my collection. Many of them are amulets that have been worn secretly on the persons of their possessors. I obtained them through the mere power of filthy lucre, and nothing can better illustrate that power than the mere fact that I did obtain them. Many of these objects were sold to me by women who dared not part with them while the priest or marabout was within hearing, but sent me away, to return at another time and complete the transaction. In my study of gems, I began with those of Egypt. Then I took up the Gnostic stones of the East, engraved with strange characters, which the people wore to keep away death. What I have gathered is the result of forty-six years of travel in ancient lands. In Africa I have unhooked the most revered fetich from the lintel of the doorway of a hut and carried it away with me, such is the power of gold.*

FIGURE 8-11.

*Portraits are universally favored.* Top: *Onyx cameo identified as Roman, from A.D. 500 to 600, depicting Victory driving a chariot.* Bottom: *Seventeenth-century Italian sardonyx cameo of a helmeted bust of Minerva. (The Sommerville collection, University of Pennsylvania)*

FIGURE 8-12.

*An excellent example of a hardstone cameo from the sixteenth century, portraying a Renaissance general in a helmet that has the form of a dolphin's head. (The Sommerville collection, University of Pennsylvania)*

FIGURE 8-13.

*Pyrrhus, in guise of Cupid, detaining his father Achilles. (The Sommerville collection, University of Pennsylvania)*

One genuine ancient sardonyx cameo that at some time in its existence had been exposed to fierce heat was a gem upon which Professor Sommerville heaped some of his highest praise. The cameo shows Minerva in the act of crowning Hercules with a laurel wreath. This, Sommerville reported, was purchased from Prince Demidoff in Florence on March 12, 1880, during a sale of items from the San Donati collection. Later glyptic experts corroborated the professor's belief in its genuineness.

The Walters Art Gallery in Baltimore has a number of glyptic-arts treasures, including ancient Egyptian scarabs, Etruscan bas-relief carvings, Mycenaean signet seals, Greek seals and intaglios (loose stones and mounted jewelry), Roman seals and cameos from the first century A.D., and some important Byzantine jewelry. Notable is a fourth-century massive gold Byzantine ring with leopards on either side of the shank, set with an oval nicolo intaglio. The subject is Victory offering a wreath. Rings of this type are believed to have been presents to important military figures, and their weight indicated the importance of the recipient. The Walters ring weighs 89.95 grams, nearly 3 troy ounces.

Other nicolo gems of importance include a pendant seal of the wreathed head of Julius Caesar with two attributes, a star and crook in the background, set in a gold mounting. The intaglio is from the collection of the earl of Bessborough, who later became the duke of Marlborough. It is the type of Caesar portrait that frequently is found on gems of the sixteenth and seventeenth century.

A nicolo gem from the earl of Arundel collection is an intaglio of a wolf. Although small, 1⁄16 inches in diameter, it is stylistically well executed.

Fifteenth-century agate intaglio mounted in a gold ring with strong Christian iconography is noteworthy for its full-length portrait of Sainte Barbara in classical dress with a martyr's palm in her hand. She stands beside a miniature tower with three windows, which is believed to allude to her devotion to the Holy Trinity; it was this dedication that led to her martyrdom. In the fifteenth century, wearing a talisman image of Sainte Barbara was believed to protect the wearer from sudden death.

Of the numerous cameos in the Walters Art Gallery, one of the loveliest is the double portrait bust of Hercules and Omphale. The lapis lazuli material is of the highest quality, solid dark blue with a few scattered spots of iron pyrite. Both figures wear lion skins on their heads, knotted at the shoulders. The cameo has impeccable provenance that places it formerly in the collection of the earl of Arundel and then in the collection of the duke of Marlborough, where the cameo was referred to as "A Ptolemy and his queen."

The Walters Art Gallery also provides a look at some genuine hat badges from the sixteenth century. As noted earlier, hat badges, or enseignes, were gems carved in bas-relief, frequently enameled, set in disc-shaped gold mountings, and drilled with a hole or fitted with a pin for attachment to the brim of a hat or beret. They were popular men's accessories in the fifteenth and sixteenth centuries.

The British Museum in London is home to the Richard Payne-Knight collection of cameos. The collection includes a number of celebrated fakes. A brown onyx head of Jupiter, supposedly by the Greek gem engraver Dioskourides and said to have been dug up in Rome in 1576, was actually carved in Italy in the late eighteenth or early nineteenth century and given a highly desirable fake signature. A white onyx Jupiter, said to have been found in a marsh, is in pristine condition, which should have warned away Payne-Knight, according to scholars. Also in the group is the famous Pistrucci head of Flora that was fraudulently sold by the unscrupulous Italian dealer Bonelli.

The Deutsches Edelsteinmuseum in Idar-Oberstein, Germany, is a unique museum, one of the finest of its type in the world. Housed in a high-rise building, the exhibit areas are on two floors; lighting is excellent, and there is ample room for displays, exhibits, and viewers. Originally the museum was devoted to the cutting and carving of gemstones, with a display of techniques as they developed in Idar-Oberstein. Later, mineral specimens and faceted gemstones were added. The first floor is devoted primarily to the evolution of the gem industry and includes early books documenting the development of cutting and carving in the city; there is also a display of materials from the mines in the surrounding mountains that were once so productive. A special area called Glyptothek, on the first floor, contains over 250 carved objects, cameos, and cameo vessels, as well as numerous works of art by contemporary master carvers (figs. 8-14, 8-15, and 8-16). The second floor also showcases some very special cameos of modern origin.

The Hermitage in Leningrad has one of the largest and best-documented collections of engraved gems to be found. The collection grew from a nucleus of stones that Peter the Great received from the Netherlands in 1721. In the second half of the eighteenth century, Catherine II purchased numerous celebrated European collections including those of: L. Natter, deBreteuil, Byres, Slade, Mengs, Lord Beverley, Louis-Philippe-Joseph (the duke of Orleans), the duke of Saint-Morys, and J. B. Casanova.

The curators of the Hermitage claim that the engraved stones in their collection provide a total picture of ancient glyptic arts and their evolution from the sixteenth century B.C. to the fourth century A.D. It is generally agreed that the museum has an unparalleled collection of engraved gems from the fifth and fourth centuries B.C.

The Kunsthistorische Museum in Vienna is home to some of the most outstanding cameo antiques and artifacts ever assembled. In this glorious building, the cabinets from the Hapsburg imperial court have been combined with many private collections to produce one of the leading museums in the world.

The museum houses the Gemma Augustea, discussed in chapter 1,

FIGURE 8-14.

*Cameos of modern origin as well as antique works can be seen in the Deutsches Edelsteinmuseum in Idar-Oberstein.* Top: *Muses and cupids are favorite carving subjects.* Bottom: *Animals test the modeling skills of some artisans. (Photos courtesy of Achim Grimm)*

FIGURE 8-15.

*A replica of the famous Grand Camée de France by Richard H. Hahn is displayed in the Deutsches Edelsteinmuseum.*

FIGURE 8-16.

*German master carver Richard H. Hahn at work on one of his most notable creations, the reproduction of the Grand Camée de France. (Copyright F. A. Becker)*

along with a celebrated bloodstone cameo of John the Baptist, unsigned, that is dated to the eleventh or twelfth century. The piece is mentioned in the 1750 Hapsburg treasury inventory as having come from Constantinople.

An onyx cameo from the first half of the thirteenth century is given special mention in the museum guidebook as having been in the 1619 estate inventory of the Emperor Matthias. The cameo shows Poseidon as patron of the Isthmian Games.

There also are numerous other famous and renowned cameos signed by such masters as Jacopo de Trezzo, Francesco Tortorino, Alessandro Masnago, and Ottavio Miseroni. Some are in commesso, a technique combining bas-relief in a unified composition of one or more stones, decorated with gold ornaments that are generally enameled.

Museum Idar-Oberstein is located in Idar-Oberstein, Germany, near the historic church-in-the-rock. Separate from the Deutsches Edelstein-museum also in that city, it is filled with carved cameos, bas-relief vessels, and very large mineral specimens. Several cabinets on the basement level of the museum are filled with cameos, reflecting the artistic talents of dozens of local master carvers. Original cutting tools are on display, along with the large stone waterwheels, the interior of an old agate-grinding mill, and a goldsmith shop.

Staatliche Munzsammlung in Munich has a state collection of over five thousand cameos in hardstone and shell; many bear signatures of celebrated carvers, such as Pistrucci. Some of the gems came from the Johann Wilhelm collection and others from the Bavarian royal family Wittelsbach.

Cameos displayed in museums continue to charm us with their designs, artistic virtuosities, and beauty of materials. Although many museums display only a small fraction of their cameo inventory, it is believed that with the upsurge of interest in the glyptic arts, curators and directors will recognize the merits of mounting new exhibits. Having more collections on public view will not only generate collecting excitement and desire in a new generation of afficionados, but will advance a full-scale revival of this historic art form.

# APPENDIX

---

#### AUCTIONS

**Auction Houses and Galleries**

Butterfield and Butterfield
220 San Bruno Avenue
San Francisco, CA 94103

Christie's
502 Park Avenue
New York, NY 10022

William Doyle Galleries
175 East 87th Street
New York, NY 10128

Dunning's Auction Service, Inc.
755 Church Road
Elgin, IL 60123

Robert C. Eldred, Inc.
1483 Main Street
Route 6A
East Dennis, MA 02641

Willis Henry Auctions, Inc.
22 Main Street
Marshfield, MA 02050

Leslie Hindman Auctioneers
215 West Ohio Street
Chicago, IL 60610

Hanzel Galleries, Inc.
1120 South Michigan Avenue
Chicago, IL 60605

Phillips Ltd.
406 East 79th Street
New York, NY 10021

Selkirk Galleries
4166 Olive Street
Saint Louis, MO 63108

Robert W. Skinner Galleries
#2 Newbury Street
Boston, MA 02116

C. G. Sloan and Company
4920 Wyaconda Road
North Bethesda, MD 20852

Sotheby's
1334 York Avenue
New York, NY 10021

A. Weschler and Son
905-9 E Street NW
Washington, DC 20004

Wolf Auction Gallery
13015 Larchmere Boulevard
Shaker Heights, OH 44120

**Auction Catalogs**

Catalogues Unlimited
P.O. Box 327
High Falls, NY 12440

The Auction Catalog Company
634 Fifth Avenue
San Rafael, CA 94901

**Subscription Newsletter of Auction Results**

Auction Forum Limited
341 West 12th Street
New York, NY 10014

**GLYPTIC-ARTS BOOKSELLERS**

The Book Block
8 Loughlin Avenue
Cos Cob, CN 06807

California Book Auction
965 Mission Street
Suite 730
San Francisco, CA 94103

Derek Content, Inc.
Crow Hill
Houlton, ME 04730

The Gemmary
P.O. Box 816
Redondo Beach, CA 90277

Harmer Johnson Books, Ltd.
21 East 65th Street
New York, NY 10021

Peri Lithon Books
P.O. Box 9996
San Diego, CA 92109

Swann Galleries, Inc.
104 East 25th Street
New York, NY 10010

Twelfth Street Bookseller
P.O. Box 3103
Santa Monica, CA 90403

**APPRAISAL ASSOCIATIONS**

The American Society of Appraisers
P.O. Box 17265
Washington, D.C. 20041

The Appraisers Association of America
60 East 42d Street
New York, NY 10165

**RETAIL SOURCES FOR CAMEOS IN THE UNITED STATES**

Barakat
429 North Rodeo Drive
Beverly Hills, CA 90210
*Antique and archaic cameos, intaglios, seals, and amulets*

George C. Houston, Inc.
550 South Hill Street, Suite 1478
Los Angeles, CA 90013
*Cameos in hard stone and shell, early twentieth century and modern*

Kim E. Hurlbert
Timeless Gem Designs
607 South Hill Street, #700
Los Angeles, CA 90014
*Custom orders for German hand-carved cameos; variety of cameo materials including opal; contemporary carvings*

Joel L. Malter and Company, Inc.
16661 Ventura Boulevard, Suite 518
P.O. Box 777
Encino, CA 91316
*Antique and modern cameos and intaglios*

Ken and Elaine Roberts
Harbour Galleries
P.O. Box 229
253 Main Street
East Greenwich, RI 02818
*Hard-stone and shell cameos, early twentieth century and modern*

# CHRONOLOGY OF ENGRAVED STONES AND CAMEOS IN EUROPE

---

**Mycenaean to Late Helladic** **1580–1100 B.C.**

Popular materials for engraved stones: chalcedony, agate, and steatite. Popular subjects: animals, flora, sea creatures, all highly stylized.

**Etruscan** **500–300 B.C.**

Etruscan craftsmen were outstanding goldsmiths and mastered the techniques of filigree and granulation. Animals and flowers were popular design subjects. Etruscan intaglios (seals) were cut in scaraboid form, with a tendency toward milled (small strokes set close together), granulated (like a string of beads), or guilloche (resembling a twisted cable) borders. (Period of Oriental influence)

**Archaic Greek** **560–480 B.C.**

Trend to naturalize the human figure and animals, but the style is excessively rigid. An era of rapid development of jewelry, and invention of coinage.

**Classic Greek** **480–330 B.C.**

Transitional time from the Archaic to the Classical in subject and design style. Subjects show more vitality; most popular are birds, animals, cupids, and winged female figures.

**Hellenistic Greek** **300–200 B.C.**

The age of Alexander the Great. Introduction of the cameo as an art form. Engraved gems carved in overly sensuous styles, with an exaggerated softness in the figural execution of both men and women. Most popular subjects: myths, portraits, gods and goddesses.

**Roman** **ca. 200 B.C.–A.D. 476**

Roman republic traditionally dates from about 500 B.C. to 27 B.C.; overseas expansion begun in third century B.C., empire formed by Octavian, called Augustus, in 27 B.C.; capital of empire moved to Constantinople in A.D. 330, marking late empire; barbarian invasions caused empire to fall in A.D. 476. Greek artisans worked for Romans and taught the art of carving and engraving. The art of cameo carving flourished, reaching its zenith in A.D. 70. Favorite subjects were gods, goddesses, heroes, mythological animals, historic scenes, and existing statues and paintings. After the fall of the Roman Empire, there was a decline in the artistic skills of workmen in all the arts.

**Holy Roman Empire** **800–ca.1250**

Established by Charlemagne in A.D. 800; although continued in some form until the nineteenth century, its greatest power lasted only until the middle of the eleventh century, after which a struggle with the papacy for dominance and the elevation of princedoms weakened it considerably. Its establishment by Charlemagne denotes the end of the ancient period in history and the beginning of medieval times, the Middle Ages. It was marked by little need or desire for art or bodily ornamentation, except for the patronage of the church and nobility. Simple luxuries were unknown. The art of cameo engraving declined to its lowest point in history.

**Renaissance** **1400–1600**

The revival of learning, facilitated by the invention of the printing press by Gutenberg in the mid-1400s. This was the age of enlightenment, of Michelangelo, Leonardo da Vinci, and Raphael. It saw the rebirth of engraving and gem carving, with royal patronage and a better standard of living for the populace contributing to the demand. Trade and manufacture made people prosperous and brought leisure and a taste for luxury and refinements of life, including collecting carved gemstones. In the Near East, carvings and engravings retain their superstitous connotations.

**Seventeenth Century**

A new and important cycle of ancient and contemporary gem collecting. As the Renaissance led to the introduction of Latin and Greek in schools, a classical education became the chief mark of distinction of a cultured person. Carved gems with classical subjects were greatly desired and enthusiastically sought out by collectors.

### Eighteenth Century

Napoléon adds to the quickened desire for cameo collecting late in the century by his public advocacy of the art; a cameo-carving school is developed in Paris. A new cycle of collecting ancient and contemporary cameos becomes a mania.

### Nineteenth Century

Cameo collecting remains very popular, until scandals and charges of fakes and forgeries discredit many nineteenth-century carvers and engravers. The reproduction of cameos and intaglios in glass causes interest in cameo collecting as an art to decline further. There is an upsurge in their use in jewelry.

### Twentieth Century

Cameos remain popular in the United States but the contemporary ones are perceived as costume jewelry. The ultrasonic carving machine introduced in the mid-1970s leads to mass production of hardstone cameos and the decline of individual skills and artistry of cameos. But hand-carved and individually produced hardstone and shell cameos once again gain importance as collectible works of art.

# GLOSSARY OF GLYPTIC-ART AND JEWELRY TERMS

**ABRASION** A type of erosion caused by friction, rubbing, or scraping of the cameo.

**ADAPTATION** An emulation of a particular kind of carving or workmanship. It may be true to the original, or it may take liberties with the original style.

**AEGIS** Shield, sometimes of goatskin.

**AGORA** Greek marketplace.

**ALLEGORY** In art, the symbolic representation of a subject.

**AMPHORA** A greek jar or vase having an egg-shaped body and a narrow cylindrical neck, usually with two handles joined to the body at the neck and shoulder of the vase.

**ANTIQUE** An artifact at least one hundred years old.

**ARCHAIC** Of an early style, as Greek cameos of the seventh and eighth centuries B.C.; or of a style that adopts the characteristics of an earlier period.

**ASKOS** Cosmetic vessel.

**ATTRIBUTE** An object closely associated with a specific person or deity.

**BARBITON** An ancient Greek musical instrument, resembling a lyre.

**BASALT** A fine vitreous black material invented by Wedgwood in 1768, having nearly the same properties as natural basalts, the igneous rock for which it was named.

**BAS-RELIEF** Sculptural relief raised from a background, in which the projection is slight and no part is undercut.

**BIGA** Two-wheeled chariot.

**BURIN** A pointed steel cutting tool.

**BYZANTINE** Relating to the style characteristics in the fifth to sixth centuries in the Byzantine empire.

**CABOCHON** A stone cut with a domed or convex top, which is unfaceted and smoothly polished; the back or base is usually flat and unpolished.

**CAMEO** A portrait or scene carved in bas-relief; usually a layer of contrasting color serves as the background.

**CANNETILLE** Open-coiled wirework technique popular in the first half of the nineteenth century.

**CAST** A method of reproducing an object by using plaster of paris to make a mold of the original piece.

**CHEVEE** (Sheh-vey) A flat gem with a smooth concave depression, with a raised (cameo) figure in the center of the depression.

**CHITON** A Greek shirt, usually a rectangle of cloth wrapped around the body and secured at the shoulders.

**CHLAMYS** A Greek cloak.

**CINQUECENTO** The 1500s, used in connection with artists from that period.

**CIRCA DATE** The approximate date of origin.

**CLASSIC** Greek art of the fifth century B.C.

**CLASSICAL** Describing ancient Greek and Roman art, from 500 B.C. to A.D. 400.

**COMMESSO** A carved cameo of the Renaissance using a combination of hardstones; subjects are usually figures or portraits; many are enameled.

**COMPOSITION** The arrangement of form, color, and line in any work of art.

**CUIRASS** Ancient close-fitting body armor.

**CUNEIFORM** Wedge-shaped writing of the Sumerians, Assyro-Babylonians, and other ancient Near Eastern peoples.

**CUVETTE** (Koo-vet) A flat gem with a smooth concave depression and a cameo in the center; interchangeable with *chevee*.

**DALMATIC** A Roman tunic, worn belted or unbelted.

**DEXTER** Situated to the right of an artifact's subject, from the subject's (not the viewer's) viewpoint. See also *sinister.*

**DIADEM** An ornamental headband worn as a badge of royalty; also worn as an emblem of regal power or dignity.

**DIPTYCH** Two panels, often hinged together, designed to close like a book, on which images are painted or carved.

**ENGRAVING** Art created by incision or carving, such as an intaglio or cameo.

**ENSEIGNE** An ornament resembling a badge that is attached to a cap or hat.

**FAÇADE** The front of a building or structure

**FAKE** An object altered or added to for the purpose of deception; a swindle or trick, or the person conducting it (Funk & Wagnalls); a counterfeit or imitation presented as genuine with fraudulent intent (Webster's New Collegiate).

**FIBULA** A clasp, like a safety pin or brooch, used to hold material together.

**FILLET** A band of fabric or leather tied around the head.

**FORESHORTENING** A method of representing objects as if seen at an angle and receding or projecting into space; not in a frontal or profile view.

**FORGERY** An object made to stimulate something of value for the purpose of deception.

**FORUM** As used in this book, a Roman marketplace.

**FREESTANDING** A work in the round, in three dimensions, not attached to a background or in relief.

**FRESCO** A painting done with pigment suspended in water on moist lime plaster.

**FRIEZE** A decorative molding, sculpture in relief, or painting in classical architecture.

**GRECO-ROMAN** Having characteristics that are partly Greek and partly Roman. Generally referring to the classical era.

**HABILLE** A cameo in which jewelry is affixed to the carved portrait, e.g., necklace, earrings, diadem.

**HIERATIC** Describing certain styles of art, in which the type of representation is fixed by religious tradition.

**HIEROGLYPHICS** The pictorial characters used for writing by the ancient Egyptians.

**HIGH RELIEF** A carving that projects so far from the background as to be nearly three dimensional or in the round.

**HIMATION** A garment worn by ancient Greek men and women, usually a rectangular cloth, 12 to 18 feet long and 4 to 6 feet wide, draped over the left shoulder and about the body, leaving the right arm free.

**ILLUMINATION** Manuscript painting in gold, silver, or brilliant colors, or a manuscript that contains cameos affixed to the cover or separate pages.

**INTAGLIO** A carved gem in which the design is engraved into the stone, below its surface, so that an impression made from the engraving is an image in relief.

**KITHARA** An ancient Greek musical instrument, similar to a harp.

**KYLIX** A Greek drinking cup shaped like a shallow bowl with two horizontal handles, sometimes on a stem.

**LACERNA** A lightweight Roman cloak, clasped at the right shoulder or in the front; similar to the Greek *chlamys* but rounded at the two corners.

**LAINA** A double toga.

**LEKYTHOS** A Greek Classic-period vase with a cylindrical body that curves inward at the foot, a tall, thin neck, and one curved handle.

**LOUPE** A small magnifying glass used by jewelers and watchmakers.

**LUTE** A guitarlike instrument, important in the sixteenth century.

**LYRE** An ancient harplike instrument, often seen as an attribute of Apollo.

**MACULATION** Spotting or stains.

**MEDIUM** As used in art, the material used for artistic expression, e.g., paint, stone, wax.

**MODELING** The means by which a three-dimensional form is suggested two dimensionally.

**MOTIF** A recurring element in an artifact that is related to theme, pattern, or color.

**MOVEMENT** In art, the sensation of action.

**NEOCLASSICAL** Having the characteristics of a style of artistic expression that is related to or adapted from the classical style.

**NIMBUS** An indication of a radiant light or aura around or above the head of a venerated person.

**OEUVRE** An artist's entire body of work.

**PALLA** A rectangular wrap used as a shawl and head covering by Roman women.

**PALUDAMENTUM** A semicircular cape worn by Byzantine men.

**PARURE** A set of jewelry made of the same type of gemstones and intended to be worn at the same time, such as a necklace, bracelet, brooch, and earrings. The modern name is suite.

**PROVENANCE** Documented history of an item, including origin and important owners.

**QUADRIGA** A chariot pulled by four horses abreast.

**REBUS** Words represented by pictures of objects or by symbols, the names of which, when sounded in sequence, reveal a message. Often used in seals.

**RELIEF** Forms that project from a background to which they remain attached. Relief may be carved or modeled shallowly, as in low or bas-relief, or deeply to produce high relief.

**REPLICA** A line-for-line copy. As much as possible, the producer uses the same materials and methods of manufacture employed in the original. A true copy.

**REPOUSSÉ** A modeled design that is hammered and punched from the back of a metal plate to raise the design on the front.

**REPRESENTATIONAL** Describing the portrayal of an object in a recognizable form.

**REPRODUCTION** As distinguished from *fake* and *forgery*, an honest copy of an antique, not primarily intended to deceive. In some designs liberties may be taken in workmanship or type of material. The artifact may be a reproduction in the style of a particular carver, artisan, or group.

**SCULPTURE** Creation of a three-dimensional form, generally in a solid material, by carving, modeling, or welding.

**SINISTER** Situated to the left of an artifact's subject, from the subject's (not the viewer's) viewpoint. See also *dexter*.

**SKYPHOS** An ancient Greek drinking vessel with a deep body, flat bottom, and two small horizontal handles near the rim.

**STELE** A commemorative slab with inscriptions, bas-relief carving, or both.

**TECHNIQUE** The method, and often the medium, used by an artist.

**TEXTURE** The surface characteristics of a work's medium, such as paint or stone, or the simulation of the subject's surface characteristics, such as for drapery or skin.

**THYRSUS** A wand or reed topped by ivy leaves, generally carried by maenads, satyrs, or the god Dionysus.

**TOGA** A semicircular draped garment worn by Roman citizens.

**TRIPTYCH** Three panels, often hinged together, on which related images are painted or carved. Sometimes the center panel is larger than the two side panels; occasionally the two side panels can be folded to cover the center panel.

**TUNIC** A Byzantine sleeved garment cut to knee or ankle length; a Roman tunic was a rectangular garment, worn at knee or ankle length, often belted.

# SUGGESTED READINGS AND BIBLIOGRAPHY


Anderson, Frank J. 1981. *Riches of the Earth.* New York: Windward Publishing Company.

Babelon, M. Ernest. 1894. *La Gravure en Pierres Fines.* Paris: G. Cres and Company.

———. 1897. *Catalogue des Camées Antiques et Modernes de la Bibliothèque Nationale.* Paris: Ernest Leroux.

Ball, Sydney H. 1950. *Roman Book on Precious Stones.* Los Angeles: Gemological Institute of America.

Barnet, Sylvan. 1985. *A Short Guide to Writing About Art.* 2d. ed. Boston: Little, Brown and Company.

Berry, Burton Y. 1968. *Ancient Gems from the Collection of Burton Y. Berry.* Bloomington, Indiana: Indiana University Art Museum.

Billing, Archibald. 1875. *The Science of Gems, Jewels, Coins, and Medals.* London. Daldy, Isbister and Company.

Boardman, John. 1968. *Engraved Gems: The Ionides Collection.* Evanston, Illinois: Northwestern University Press.

Bohr, R. L. 1968. *Classical Art.* Dubuque, Iowa: William C. Brown Company.

Booth, A. 1890. *Gems, Cameos and Amber.* Gloucester, England: John Bellows.

Campbell, Joseph. 1988. *The Power of Myth.* New York: Doubleday.

Charles, Russell, J. 1975. The challenge of carving shell cameos. *Lapidary Journal* 29, no. 2 (May):488-94.

Cirker, Hayward, and Blanche Cirker. 1967. *Dictionary of American Portraits.* New York: Dover Publications.

Collon, Dominique. 1987. *First Impressions.* London: British Museum Publications.

Content, Derek J., ed. 1987. *Islamic Rings and Gems.* London: Philip Wilson Publishers.

Davenport, Cyril. 1900. *Cameos.* London: Seeley and Company.

Dimitrova-Milcheva, Alexandra. 1981. *Antique Engraved Gems and Cameos.* Bulgaria: Septemvri Publishing House.

Dorn, Sylvia O'Neill. 1974. *The Insider's Guide to Antiques, Art and Collectibles.* New York: Doubleday.

Editors of McGraw-Hill. 1966-68. *McGraw-Hill Modern Men of Science.* New York: McGraw-Hill.

Fiorelli, Anna, ed. 1989. *Corals and Cameos: The Treasures of Torre del Greco.* Italy: Banca di Credito Popolare.

Foskett, Daphne. 1989. *Miniatures: Dictionary and Guide.* London: Antique Collectors Club.

Frazier, Si, and Ann Frazier. 1988. Museum Idar-Oberstein. *Lapidary Journal* 42, no. 9 (December):41-57.

———. 1990. Quartz building blocks. *Lapidary Journal* 43, no. 11 (February): 89-91.

Furtwängler, Adolf. 1900. *Die Antiken Gemmen.* 3 vols. Berlin: Greseche and Devrient.

Furuya and Company. 1984. *A Catalogue of the Collection of Engraved Gems.* Kofu, Japan, and Idar-Oberstein, Germany: Furuya and Company.

Garside, Anne, ed. 1979. *Jewelry, Ancient to Modern.* Baltimore: Trustees of the Walters Art Gallery.

Gifhetti, Romolo. 1955. Catalog accompanying the exhibition *"Gemme e Cammei delle Collezioni Comunali,"* held at Dei Musei Comunali, Rome, Italy.

Giuntoli, Stefano. 1989. *Art and History of Pompeii.* Rome, Italy. Bonechi.

Goldemberg, Rose Leiman. 1978. *Antique Jewelry: A Practical and Passionate Guide.* New York: Crown.

Goldstein, Sidney M., Leonard S. Rakow, and Juliette K. Rakow. 1982. *Cameo Glass.* Corning, New York: The Corning Museum of Glass.

Gorely, Jean. 1950. *Wedgwood.* New York: Gramercy Publishing Company.

Gray, Fred L. 1983. Engraved gems: A historical perspective. *Gems and Gemology* 19, no. 4 (Winter):191-201.

Gwinnett, A. J., and L. Gorelick. 1979. Ancient lapidary. *Expedition* 22, no. 1:17-32.

Gwinnett, A. J., and L. Gorelick. 1981. Close Work without Magnifying Lenses? *Expedition* 23, no. 2:27-34.

Hall, G. K. 1960. *Portrait Catalog.* New York: New York Academy of Medicine.

Hamilton, Edith, 1969. *Mythology.* Boston: Little, Brown and Company.

Havelock, Christine Mitchell. 1981. *Hellenistic Art.* Rev. ed. New York: W. W. Norton and Company.

Haynes, Colin. 1988. *The Complete Collector's Guide to Fakes and Forgeries.* Greensboro, North Carolina: Wallace-Homestead.

Hobson, Burton. 1971. *Historic Gold Coins of the World.* Garden City, New York: Doubleday.

Hogrefe, Jeffrey, ed. 1982. *The Antiques World Price Guide.* New York: Doubleday.

Howard, Margaret Ann. 1988. The delicate cameo. *Lapidary Journal* 42, no. 8 (November):63-65.

Jokelson, Paul. 1968. *Sulphides.* New York: Thomas Nelson and Sons.

Jones, Mark, ed. 1989. Catalog accompanying the exhibition "Fake? The Art of Deception," held at The British Museum, London.

Kagan, Ju. 1973. *Western European Cameos in the Hermitage Collection.* Leningrad: Aurora Art Publishers.

Kemp, Russell M. 1989. Cameos and intaglios of stone. *Lizzadro Museum* periodical, (Winter-Spring):4-20.

King, C.W. 1872. *Antique Gems and Rings.* Vols 1 and 2. London: Bell and Daldy.

———. 1885. *Handbook of Engraved Gems.* London: George Bell and Sons.

Kovel, Ralph, and Terry Kovel. 1967. *Know Your Antiques.* New York: Crown.

Kris, Ernst. 1932. *Catalogue of Postclassical Cameos in the Milton Weil Collection.* Vienna: Anton Schroll and Company.

Macht, Carol. 1957. *Classical Wedgwood Designs.* New York: Gramercy Publishing Company.

Matlins, Antoinette L., and A. C. Bonanno. 1987. *Jewelry and Gems: The Buying Guide.* South Woodstock, Vermont: Gemstone Press.

McConkey, Kenneth. 1989. *Edwardian Portraits: Age of Opulence.* London: Antique Collectors Club.

Middleton, J. Henry. 1891. *The Engraved Gems of Classical Times.* London: Cambridge University Press.

———. 1892. *The Lewis Collection of Gems and Rings.* London: C. J. Clay and Sons.

Miller, Anna M. 1988. *Gems and Jewelry Appraising: Techniques of Professional Practice.* New York: Van Nostrand Reinhold.

———. 1989. *Illustrated Guide to Jewelry Appraising: Antique, Period, and Modern.* New York: Van Nostrand Reinhold.

Mills, John Fitzmaurice. 1980. *How to Detect Fake Antiques.* New York: Desmond Elliott.

Murray, John. 1804. *Gems, Selected from the Antique.* London: C. Whittingham.

Nassau, Kurt. 1984. *Gemstone Enhancement.* England: Butterworths.

Natter, Lorenz. (Private Publication, 1754.) *Traité de la méthode antique de graver en pierres fines, Comparée avec la méthode moderne.* viii.

Neverov, O. 1976. *Antique Intaglios in the Hermitage Collection.* Leningrad: Aurora Art Publishers.

Oberleitner, Wolfgang. 1985. *Geschnittene Steine.* Vienna: Hermann Bohlaus.

Ogden, Jack. 1982. *Jewellery of the Ancient World.* New York: Rizzoli.

Osborne, Duffield. 1912. *Engraved Gems.* New York: Henry Holt and Company.

Pavitt, William. 1970. *The Book of Talismans.* 4th reprint. London: Stephen Austin and Sons.

Pepper, Stephen C. 1949. *Principles of Art Appreciation.* New York: Harcourt, Brace and Company.

Prendeville, James. 1841. *An Historical and Descriptive Account of the Famous Collection of Antique Gems Possessed by the Late Prince Poniatowski.* London: Henry Graves and Company.

Reilly, D. R. 1953. *Portrait Waxes.* London. B. T. Batsford.

Richter, Gisela M. A. 1920. *Catalogue of Engraved Gems of the Classical Style.* New York: Metropolitan Museum of Art.

———. 1942. *Ancient Gems from the Evans and Beaty Collection.* New York: Metropolitan Museum of Art.

———. 1956. *Catalogue of Engraved Gems: Greek, Etruscan, and Roman.* Rome: L'Erma di Bretschneider, for the Metropolitan Museum of Art.

———. 1971. *Engraved Gems of the Romans.* New York: Phaidon Press.

Riley, Olive L. 1963. *Your Art Heritage.* New York: McGraw-Hill.

Ritchie, I. A. 1970. *Carving Shells and Cameos.* New York: A. S. Barnes and Company.

St.-Memin. 1862. *The St.-Memin Collection of Portraits.* New York: E. Dexter.

Savage, George. 1968. *The Antique Collector's Handbook.* London: The Hamlyn Publishing Group.

Scarisbrick, Diana. 1984. *Jewellery.* London: B. T. Batsford.

Sedgwick, Paulita. 1974. *Mythological Creatures.* New York: Holt, Rinehart and Winston.

Sines, George, and Yannis A. Sakellarakis. 1987. Lenses in antiquity. *American Journal of Archaeology,* no. 91:191-96.

Sinkankas, John. 1968. *Van Nostrand's Standard Catalog of Gems.* New York: Van Nostrand Reinhold.

———. 1988. *Gemstone and Mineral Data Book.* Prescott, Arizona: Geoscience Press.

Siviero, Rodolfo. 1959. *Jewelry and Amber of Italy.* Naples, Italy: Museo Nazionale.

Smith, H. Clifford. 1973. *Jewellery.* Reprint. London: Metheun and Company.

Sommerville, Maxwell. 1889. *Engraved Gems: Their History and Place in Art.* Philadelphia: Sommerville.

———. 1901. *Engraved Gems.* Philadelphia: Drexel Biddle.

Strong, Roy C. 1969. *The English Icon: Elizabethan and Jacobean Portraits.* New York: Pantheon.

———. 1969. *Tudor and Jacobean Portraits.* 2 vols. London: H. M. S. O.

Tagliamonte, Nino. 1984. History of shell cameos. *Lapidary Journal* 42, no. 8 (November):22-32.

Vermeule, Cornelius C. 1956. *Cameo and Intaglio: Engraved Gems from the Sommerville Collection.* Philadelphia: University of Pennsylvania, The University Museum.

Westropp, Hodder M. 1867. *Handbook of Archaeology.* London: Bell and Daldy.

Willems, J. Daniel. 1972. Cameos. *Mineral Digest* 2 (Summer):42-51.

Wills, Geoffrey. 1989. *Wedgwood.* Secaucus, New Jersey: Chartwell Books.

Zeitner, June Culp. 1988. The fine art of gem engraving. *Lapidary Journal* 42, no. 8 (November):22-32.

# INDEX